Birding Hot Spots of Santa Fe, Taos, and Northern New Mexico

NUMBER FIFTY-ONE

W. L. Moody Jr. Natural History Series

A⫪M nature guides

Sage Thrasher

Birding Hot Spots
of Santa Fe,
Taos, and
Northern New Mexico

Judy Liddell & Barbara Hussey

TEXAS A&M UNIVERSITY PRESS ▶ COLLEGE STATION, TEXAS

This paper meets the requirements of
ANSI/NISO Z39.48–1992 (Permanence of Paper).
Binding materials have been chosen for durability.

LIBRARY OF CONGRESS CATALOGING-IN-PUBLICATION DATA

Liddell, Judy, 1942– author.
  Birding hot spots of Santa Fe, Taos, and northern New Mexico / Judy Liddell and Barbara Hussey.—First edition.
      pages cm—(W.L. Moody Jr. natural history series ; Number fifty-one)
  Includes bibliographical references and index.
  ISBN 978-1-62349-254-0 (flexbound : alk. paper)—
  ISBN 978-1-62349-258-8 (e-book)
  1. Birds—New Mexico—Santa Fe—Guidebooks. 2. Bird watching—New Mexico—Santa Fe—Guidebooks. 3. Birds—New Mexico—Taos—Guidebooks. 4. Bird watching—New Mexico—Taos—Guidebooks. 5. Birds—New Mexico—Guidebooks. 6. Bird watching—New Mexico—Guidebooks. I. Hussey, Barbara, 1946- author. II. Title. III. Series: W.L. Moody Jr. natural history series ; no. 51.
  QL684.N6L45 2015
  598.072'3478956—dc23
                                    2014039319

Photographs by Judy Liddell unless otherwise indicated.
Cover photo by Lou Feltz.

▶ This book is dedicated to Audubon New Mexico,
whose mission is to conserve and restore New Mexico's
natural ecosystems and the earth's biological diversity,
focusing on birds, wildlife, and habitats
for the benefit of humanity.

Contents

Preface

As we wrote *Birding Hot Spots of Central New Mexico* and talked about what sites to include, we laughed as we decided not to include the Cochiti Lake area and said, "That will be for our next book." We never thought any more about this conversation until the book was being launched and we were asked, "What will your next book be?" North-central New Mexico was a natural as it encompasses many of our favorite locations.

We spent 2012 revisiting areas we wanted to include in the book, gathering site data, meeting managers and wildlife biologists at each location, and connecting with local birders. Birders, as always, are generous in sharing information and often accompanied us to their favorite locations. At the request of current readers, we added information to each site indicating whether pets were welcome and the location of nearby camping, if not available on-site.

Like our first book, this book is site based, providing a more in-depth look at specific areas, and should be considered a supplement to the New Mexico Ornithological Society's excellent *New Mexico Bird Finding Guide,* which covers the entire state.

We are grateful to all of the birders, both local and visiting, who have entered their sightings into Cornell Laboratory of Ornithology's eBird database. That information has provided the basis for the information about species that can be seen at each site and has enabled us to provide reliable information for the annotated checklist.

When I was in high school and living in Southern California, my family visited north-central New Mexico. One night we pulled our travel trailer into the Ghost Ranch, which had been open only a year, and were invited to spend the night in the Ghost House. I fell in love with the imposing red-rock cliffs that contrasted with the monsoon-greened vegetation. It was an image that I held for the next 15 years until I visited New Mexico again. My sister had married a New Mexico native, and I visited with my family every few years—

always including areas in the north. After I moved to Albuquerque in 1994, I set out to explore these areas in more depth. There is something about high mountain meadows and alpine forests that touches my soul.

—Judy Liddell

As I begin my fortieth year as a birder, I recall with fondness some of the remarkable times spent birding and just enjoying the wonders of northern New Mexico. From our home in Albuquerque, my family and I made countless trips north for skiing, bicycling, camping with my daughter Linda's Girl Scout troop, and sleeping under a sky full of stars. We spent time hiking, fly fishing, gold panning, and exploring ruins and ghost towns. There were Pueblo dances, operas in Santa Fe, murder mystery weekends in Cimarron, train rides from Chama, and conferences at Ghost Ranch and Los Alamos. We bought apples and raspberries and dined on chile in restaurants housed in ancient adobes built before New Mexico was part of the United States. All these memorable adventures were framed by the opportunity to discover the incredible birdlife of northern New Mexico.

After several years experiencing the excitement of urban life in our nation's capital, my husband and I have returned to New Mexico. Our first granddaughter was born here not long ago. It was time to come home.

—Barbara Hussey

Acknowledgments

This guide has been enriched through the generous contributions of numerous New Mexico birders and natural history enthusiasts who shared their wealth of information about local sites and species and reviewed site descriptions.

Several people went the extra mile, taking us birding with them in their communities and sharing their favorite locations. Cheryl Grindle accompanied us on site visits, lent us trail books, and helped us network with other birders. Mary Jo Kelly drove to Eagle Nest from her home near Questa to show us her favorite birding locations in Cimarron Canyon State Park. In addition to going birding with us, Jan Saunders and Richard Dickerson introduced us to local resources in the Angel Fire area. Jim and Ann Ellen Tuomey met us on two different days in Taos and at the Enchanted Cross Country Ski Area to bird and provide critical local information. Joe and Sally Fitzgibbon invited us to their home on more than one occasion to enjoy the multitude of birds that visit their feeders; Joe was our tour guide in Los Alamos and assisted our connecting with other Los Alamos birders. In addition to Steve Fettig's being a wildlife biologist for Bandelier National Monument, we benefited from his expertise as a teacher and mentor to other birders in Los Alamos and his dedication to banding birds at the Valles Caldera National Preserve. We could not have completed the book without the support of these generous individuals.

Some individuals were site angels—collecting detailed information on sites where we were not physically able to travel—Ken Cole and Steve West kept detailed records of birds seen along the Rio Chama on a rafting trip, as did Lottie Hays. Cole Wolf wrote the details of his Ptarmigan Trek so we could include that specialized site.

Thanks also go to the following individuals: Collin Adams, Jonathan Batkin, Bill Burk, Roger Clark, Steve Cox, Wyatt Egelhoff, Suzanne Fahey, Sylvia Fee, Bernie Foy, Siscily Fraley, Jerry Friedman, Rebecca Gracey, Kathy Granillo, Dave Hawkesworth, Linda Heinze, Bill Howe, William Jaremko-

Wright, Linda Mowbray, Jim O'Donnell, Meg Peterson, Nick Peterson, Marty Pfeiffer, Cassidy Ruge, Graziella Singleton, Dale Stahlecker, Tom Taylor, Sarah Turner, Ernie and Jan Villescas, Elizabeth Winter, and Dave Yeamans.

We owe our deep gratitude to Christiana Burk, who generously lent her architectural skills and many hours of labor to create the trail and driving diagrams.

The book has been enhanced by the generosity of several talented photographer-birders: Warren Berg, Lou Feltz, Bernie Foy, Nancy Hetrick, Jim Joseph, Sally King, Joe Schelling, Jim Tuomey, Mouser Williams, and Cole Wolf.

Christopher Rustay always responded to questions about sites and species. Brian Millsap and Kristin Madden provided raptor information. Baker H. Morrow, fellow of the American Society of Landscape Architects, was able to assist in locating information on habitat vegetation. John Fleck, *Albuquerque Journal* science writer, provided explanations about water flows along the Rio Chama and a wealth of scientific information about the effects of climate change on forests in north-central New Mexico. Gale Garber, executive director of HawksAloft, shared monitoring data for the upper Rio Grande.

Thanks are due to all of the diligent birders who have entered their north-central New Mexico sightings, including numbers, into eBird. These sightings assured the accuracy of the individual site descriptions, as well as the annotated checklist.

We are grateful for the assistance and review from several staff at Randall Davey Audubon Center: Karyn Stockdale, executive vice president and executive director; Carol Beidelman, director of bird conservation; and Steven J. Cary, operations and resource manager.

We are appreciative of the review assistance of Robert Findling, director of land protection and stewardship, The Nature Conservancy, and to Jeane Dugan, who coordinated input and review from Ghost Ranch.

Arch Wells, chief forester (retired), Bureau of Indian Affairs, provided technical input into the section on habitats.

Thanks go to Kai-t Blue Sky, wildlife biologist at Pueblo de Cochiti, for meeting with us and providing tribal input for our Cochiti area sites. We are also grateful to a number of the Natural Resources staff at Ohkay Owingeh Pueblo for meeting with us and approving our site description: Naomi Archuleta, Connie Martinez, and Larry Phillips. In addition, Charlie Marcus, former manager of Ohkay Owingeh Fishing Lakes (Tsay Corporation) and Charles Lujan, former natural resources director, met with us on-site and answered many questions.

The following individuals from the State Parks Division of the New Mexico Energy, Minerals and Natural Resources Department met with us to provide information about their particular park: Ramon Gallegos, Villanueva State Park; Daniel Gurule, Cimarron Canyon State Park; Katherine Germain, Vietnam Veterans Memorial State Park; Paul Lisko, Fenton Lake State Park; and Christopher Vigil and Daniel Alcun, Coyote Creek State Park. Beth Wojahn, senior editor and communications specialist for New Mexico Energy, Minerals and Natural Resources Department, and David Certain, biologist and state trails coordinator, reviewed all state park site descriptions.

We received assistance and review from wildlife biologists Esther Nelson, Mary Orr, and Jo Wargo from Santa Fe National Forest and Francisco Cortez and Marty Pfeiffer from Carson National Forest.

Sally King, park ranger, and Steve Fettig, wildlife biologist, provided a wealth of information about the birds and resources of Bandelier National Monument. Roger Clark, park ranger, reviewed our information for Pecos National Historical Park.

The following individuals from the Bureau of Land Management provided information and review: Jackie Leyba, Gregory Gustina, John Bailey, and Valerie Williams.

Clint Henson, northeast area public information specialist, coordinated input on a number of the New Mexico Department of Game and Fish Wildlife Areas. Don Arevalo provided information for the Red River Fish Hatchery, and Kristen Madden, bird program manager, provided information on bird species. Ross Morgan, northwest region information and education officer, and Mark Bundren, Chama district officer, provided input on sites along the Rio Chama.

We are grateful for the assistance of Rob Larrañaga, refuge manager, Las Vegas National Wildlife, for meeting with us and coordinating review from Phillip Garcia, wildlife biologist, and Debbie Pike, visitor services manager. Leann Wilkins, manager at Maxwell National Wildlife Refuge, was an ongoing source of information.

We are grateful for the prompt response from Kimberly DeVall, interpretation and education specialist, and Bob Parmenter, director of science and education, with Valles Caldera National Preserve.

Staff from the Army Corps of Engineers provided assistance and review: John Mueller, lead natural resource specialist for Abiquiu Lake; and Marc Rosacker, project manager, and Nicholas Parks, park ranger, at Cochiti Lake.

Gilbert Martinez, Santa Fe Parks Department, and Leroy Pacheco, Taos Parks Department, provided review of park sites in their communities. Vermejo Conservancy District reviewed site information for Stubblefield Lake. Ski Santa Fe reviewed our description for the Santa Fe Ski Area.

Our thanks go to the staff of Texas A&M University Press, especially Shannon Davis, editor-in-chief, for having faith in us to produce a second book. While we did not require as much hand holding as the first time around, we are grateful for the expert guidance and support throughout the editorial and production process from Pat Clabaugh, associate editor, and to Cynthia Lindlof, our copyeditor.

We apologize if we have unintentionally omitted anyone.

Birding Hot Spots of Santa Fe, Taos, and Northern New Mexico

CHAPTER 1

North-Central New Mexico's Geography and Habitats

North-central New Mexico is a lure for those who love the outdoors. It has a number of nationally known tourist destinations and boasts a list of 276 species of birds seen on a regular basis, including Bald and Golden Eagles, Lewis's Woodpecker, Pinyon Jay, Evening Grosbeak, and Dusky Grouse. The described hot spots are clustered into eight geographic areas: greater Santa Fe area, greater Taos area, Enchanted Circle Scenic Byway and nearby areas, Jemez Mountains and Los Alamos, along the Rio Chama, Cochiti Lake area, along the upper Pecos River, along I-25 north of Santa Fe, and two high-elevation locations for specialty birds. Sites covered in this guide are located in seven counties: Sandoval, Santa Fe, Los Alamos, Taos, Rio Arriba, San Miguel, and Colfax.

Geography

The geography of north-central New Mexico is dominated by the rugged and majestic Sangre de Cristo Mountains, the southern portion of the Rocky Mountains, with seven peaks towering over 13,000 feet. Sangre de Cristo is the Spanish term for "blood of Christ," reflecting the color the mountains assume at sunset.

Rivers are also an important part of north-central New Mexico's geography. The Rio Grande, which originates in southwestern Colorado, is considered the lifeblood of New Mexico and enters the state north of Taos, where it meanders through the deep canyons of the Rio Grande Gorge, the Española Valley, and south through Cochiti Pueblo. The Rio Chama, the third-largest tributary of the Rio Grande, also originates in southern Colorado

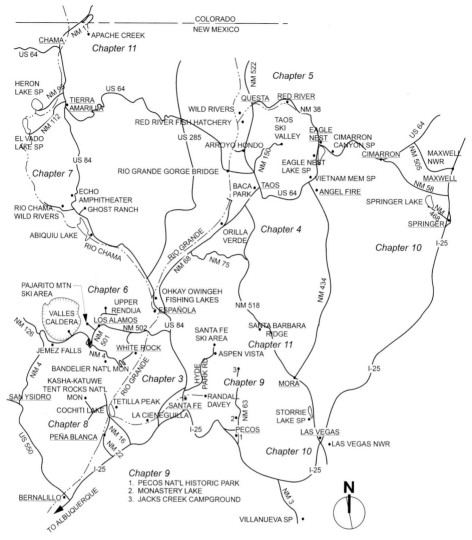

Map 1. Sites in North-Central New Mexico

and winds through New Mexico along the eastern edge of the Colorado Plateau through sandstone canyons and Abiquiu Lake, before entering the Española Valley, where it joins the Rio Grande at the location of the first European settlement in the Southwest. The Pecos River, also a tributary of the Rio Grande, does not flow into the Rio Grande until after it reaches Texas. Its headwaters are in the Pecos Wilderness high country in the Sangre de Cristos. It flows through the upper Pecos River Valley, crosses Interstate 25, and then snakes through the Villanueva State Park's sandstone cliffs before entering the Great Plains.

Vegetation Zones and Habitats

Birdlife is closely connected to habitat. Various factors affect the type of vegetation and habitat throughout this area. Elevation, latitude, exposure, prevailing winds, temperature, and rainfall contribute to this diversity. For example, a vegetation type on the south side of a slope will occur at a higher elevation than on the north side. Visitors to northern New Mexico are often puzzled when they encounter habitats and bird species at different elevations than they might be used to. For instance, the plateau north of Taos contains Great Basin shrub-grassland and piñon-juniper habitats; surprisingly, the road to the Red River Fish Hatchery descends into ponderosa pine habitat, which is usually at a higher elevation than piñon-juniper.

For conservation purposes, north-central New Mexico is part of the Southern Rocky Mountains Ecoregion. A variety of terminology is used to define habitats or vegetation zones. The New Mexico Avian Conservation Partners, a collaboration of more than 14 governmental and nonprofit organizations, plus university and private researchers, recognizes the following 15 vegetation zones in north-central New Mexico, each defined by its characteristic dominant plants. Some site descriptions refer to ecotones that occur when habitats overlap and species intermingle.

- Alpine tundra: Vegetation at elevations generally over 11,500 feet is dwarfed and gnarled and often referred to as Krummholz. It is found on Santa Barbara Ridge.
- Spruce-fir forest: The dominant plant species are Engelmann spruce, blue spruce, bristlecone pine, and corkbark fir. It can be found at the Santa Fe Ski Area and Taos Ski Valley.
- Mixed conifer forest: The primary plant species are Douglas-fir, white fir, ponderosa pine, aspen, water birch, Rocky Mountain juniper, and southwestern white pine. Examples include Aspen Vista Picnic Area, Pajarito Mountain Ski Area, and Apache Creek. Some areas within mixed conifer forests contain large stands of aspens, most of which originated from fire events that decimated the conifers. Aspen is one of the first trees that emerges after a fire, sprouting from dormant rootstock. Other aspen groves border alpine meadows that were heavily grazed by livestock.

- Transition zone or ponderosa pine forest: The trees include ponderosa pine in an open forest with grassy openings, Gambel oak, western chokecherry, and New Mexican locust. Sites with this type of habitat include Black Canyon Campground along Hyde Park Road, Cimarron Canyon State Park, and Jemez Falls.
- Piñon-juniper woodland: The dominant tree/shrub/plant species are piñon pine, juniper, Apache plume, mountain mahogany, and four-wing saltbush. This habitat can be found at Rio Grande del Norte National Monument: Wild Rivers, areas along the Rio Chama Wild and Scenic River, and Kasha-Katuwe Tent Rocks National Monument.
- Montane riparian: This habitat occurs as a narrow, often dense grove of broad-leaved, deciduous trees. Holy Ghost Campground is an example of this type of riparian area.
- Montane shrub: This habitat is a patch or a strip within other more extensive types of vegetation, such as a wash, arroyo, or escarpment, where there is less available moisture than in surrounding areas. Sections of montane shrub occur along Rio Grande del Norte National Monument: Orilla Verde.
- Middle-elevation riparian: This is a tree- and/or shrub-dominated area along a river or stream, including the cottonwood *bosque* (Spanish for "woodlands"). Examples include Ohkay Owingeh Fishing Lakes along the Rio Grande and Villanueva State Park.
- Subalpine wet meadow: This habitat is a seasonally wet area at high elevations below the tree line. Examples include the meadow at Valles Caldera, Jacks Creek Campground, and areas in the Taos Ski Valley.
- Montane grassland: This habitat is found at high elevation in the upper areas of the Valles Caldera National Preserve and at Jacks Creek.
- Emergent wetlands and lakes: These include both seasonal and permanent wetlands as well as ponds and lakes. Examples occur at Maxwell National Wildlife Refuge, Monastery Lake, and Fred Baca Park.
- Great Basin shrub: This is a high-desert habitat in areas that experience snow during the winter and is dominated by a variety of sages. It can be found along Forest Road 151 near Abiquiu Lake and on the mesa north and west of Taos.

- Upland desert scrub (referred to in this guide as desert scrub): This habitat is dominated by sand sagebrush in combination with other shrubs and cacti, such as four-wing saltbush, chamisa (rabbit brush), prickly pear, and cholla cactus. While not prevalent in north-central New Mexico, it can be found along the road to the Tetilla Peak Recreation Area and near La Cieneguilla Marsh.
- Plains-mesa grassland: This type of habitat, sometimes referred to as shortgrass prairie, is primarily grasses, such as blue grama and buffalo grass, and is found near sites along I-25.
- Agricultural: This habitat includes areas where crops are planted and harvested. Sites with agricultural areas include Peña Blanca and the roads into El Vado and Heron Lake State Parks.

The vegetation in north-central New Mexico is undergoing change because of gradually warming temperatures. Several site descriptions include mention of damage or changes as a result of recent wildfires. These fire events that are the most obvious result of forest change are only part of the story. While north-central New Mexico has had alternating wet and dry periods over time, the drought that started in 2000 (and continued through the time this guide was written) has lasted longer and has had more drastic effects on forest habitats than in the past. According to climate scientists at Los Alamos National Laboratory, located in this region, forest changes are accelerating. Even during years with the same amount of rain and snowpack as the area experienced in the past, trees are still dying off because temperatures are warmer. As a result of these changes, habitats described in the guide, and their accompanying bird species, may look different in the future.

CHAPTER 2

Helpful Information

How to Use This Guide

The guide includes nine chapters devoted to distinct areas within north-central New Mexico. Each description includes a general account of the highlights of the site and recommended route; counties covered; Internet website or Facebook page, if available; target species; listing of other birds that might be seen by season; driving directions; public transportation route, if available; fees; special considerations and hazards; facility information, including information about accessibility for people with varying abilities; availability of restrooms, water, and picnic facilities; and general information on the nearest gas, food, and lodging.

Target Birds

Target birds are those that local birders have suggested are the species that visitors to an area are most interested in viewing. The seasons for each of the sites are broken down into winter (generally December–March), summer (June–August), and migration, which can overlap these two seasons. Spring migration can begin for some species toward the end of February and continue through May. Fall migration can begin as early as mid-July and finish toward the end of November, depending on the species. As you read the list of species at a specific site and find one that is a target species for you, we recommend that you consult the annotated checklist near the end of the book, where specific months are provided for arrival and departure.

eBird Hotspots

Some site descriptions list eBird Hotspots associated with it, not only to assist you in more easily checking on recent sightings but also to encourage you to enter your sightings in existing Hotspots. The Cornell Laboratory of

Ornithology and the National Audubon Society launched eBird in 2002 on the Internet to provide a way for "citizen scientists" (all birders) to contribute to the knowledge base of bird distribution and abundance and has continued to evolve. To obtain current information on eBird Hotspot sightings, click on Explore a Location on the Explore Data tab and enter the name of the Hotspot (www.ebird.org). Or use the Find Nearby Hotspots on the BirdsEye app that can be downloaded on a smart phone or tablet.

Directions

We recommend supplementing the directions in this guide with New Mexico state and local road maps available through New Mexico Tourism Department offices (http://newmexico.org/map/), local tourist information centers, mobile phone apps, or GPS devices. All directions to the sites in this guide originate from a city or town easily found on a state road map (Santa Fe, Albuquerque, Taos, and Las Vegas, New Mexico). Some sites have directions from multiple towns. Interstate exit numbers correspond roughly with the mileage distance on that highway from the southern or western starting point of either I-25 or I-40 in New Mexico. As individual vehicle odometers vary, all mileages are approximate. We also suggest stopping at the park, monument, or refuge visitor centers listed in this guide to obtain trail maps when available.

GAIN Permits and HMAV

A few of the sites require a New Mexico Gaining Access into Nature (GAIN) permit and Habitat Management and Access Validation (HMAV). GAIN is a program offered by the New Mexico Department of Game and Fish (NMDGF).

These permits are mandatory for birding on trails within the Colin Neblett Wildlife Management Area near Cimarron Canyon and Eagle Nest Lake State Parks (chapter 5); the Rio Chama Trail between El Vado Lake and Heron Lake State Parks (chapter 7); and Bert Clancy and Terrero fishing areas along the Pecos River on NM 68 (chapter 9). GAIN and HMAV permits are not available on-site at any of the birding locations in this guide. They must be purchased in advance at any NMDGF office, online (https://onlinesales.wildlife.state.nm.us/), or from private license vendors such as hunting and fishing supply dealers or other retail or sporting goods stores.

Annual GAIN permits are valid from April 1 to March 31. NMDGF offers a temporary permit valid for five consecutive days. Permit prices may be different for those with addresses outside New Mexico. HMAV must be purchased with GAIN, as well as with all New Mexico hunting and fishing licenses.

Federal Passes

There are several types of recreational passes issued by the federal government. All of the passes are honored nationwide at all Forest Service, National Park Service, Bureau of Land Management, Bureau of Reclamation, and US Fish and Wildlife Service sites charging entrance or standard amenity fees.

Accessibility

For those readers who have varying abilities, we have tried to describe the conditions present at each site regarding paved or dirt trails, steepness of slope, availability of hand railings and other amenities, loose gravel, exposed tree roots, and possible presence of mud, ice, and other hazards.

Water

Drinking water sources can be scarce in some areas of New Mexico, especially at sites far from towns and in desertlike areas. The availability of drinking water is listed in each site description. Consider all surface water sources to be contaminated. We recommend bringing your own water with you on hikes and in your vehicle.

Pets

At most sites in this guide, pets are permitted if they are on a leash. At a few sites, they are not allowed except inside a vehicle.

Gas, Food, and Lodging

For the convenience of users of this guide, we have listed the nearest gas, food, and lodging for each site. It is recommended to refill gas tanks often, as gas stations are often few and far between on many north-central New Mexico roads.

Special Considerations and Hazards

We have listed some of the more important special considerations and hazards birders may encounter at each site description. Many of these are self-explanatory or have brief explanations in the individual narrative. The few that need more detailed information are listed here.

Black Bears and Cougars

Black bears are a concern at many of the birding sites listed. Not all black bears in New Mexico have black fur; some are brown, cinnamon, or blond. It is important not to leave food unattended on picnic tables—even briefly. US Forest Service personnel have been known to confiscate abandoned food. Never behave like prey and run from a black bear. Back up slowly and walk away. If a bear approaches you, make a lot of noise. Remain standing upright. Never lie down to play dead or turn your back on an approaching black bear.

Do not behave like prey and run from a cougar. Cougars hunt for deer, their primary prey, at dawn and dusk and usually avoid people. Hike with a friend if you are birding early or late at the sites where they are mentioned, a good idea at all times.

Rattlesnakes

Several species of rattlesnakes are known to inhabit parts of north-central New Mexico, including prairie rattlesnake and western diamondback rattlesnake, and can occur up to altitudes of 9,500 feet or higher. Unless they feel threatened, snakes will not usually bother humans. Walk in cleared areas where it is easy to see where you step or reach with your hands. Wear sturdy hiking boots. Snakes often seek shade during intense summer heat.

If bitten, remain calm and put a safe distance between you and the snake. Call 911 for transport to a medical facility. Antivenin is the only definitive treatment. While awaiting transport, call the New Mexico Poison Center at 1–800–222–1222 for guidance. If you are out of cell phone range, get to a hospital as soon as possible.

Hunting Seasons

A few of the sites in this guide are also areas where hunting is allowed at certain times of the year. For more information on locations and seasons, visit the New Mexico Department of Game and Fish Web site (http://www.wildlife.state.nm.us/recreation/hunting/) or pick up a copy of its free publication *New Mexico Hunting Rules and Information*, available at most New Mexico state parks and visitor centers.

Plague

Present in several areas of north-central New Mexico, plague is a disease of wild rodents and rabbits caused by the bacterium *Yersinia pestis*. It is spread among animals and to humans by the bites of infected fleas. Animals that are most often infected include rock squirrels, prairie dogs, pack rats, chipmunks, rabbits, and mice. When an animal with plague dies, the infected fleas must find a new host. This may be another rodent, a pet, or a person. Avoid contact with wild rodents and their fleas, nests, and burrows. Prevent pets from hunting by keeping them on a leash while hiking.

Poison Ivy

Common in New Mexico, the western species (*Toxicodendron rydbergii*) is the nonclimbing, bushy variety. It can be recognized by its sometimes shiny

Western species of poison ivy in fall

leaves in groups of three. It is known to be present at nearly half of the locations featured in this guide. The leaves of poison ivy often change to an attractive scarlet color in the fall, resembling Virginia creeper.

Washing with soap and water immediately after contact may remove most of the toxic oil (urushiol). Its residue on an unleashed dog's fur can convey this oil to humans. Symptoms of exposure may not appear for several hours but may last for a week.

Stinging Nettle

Usually occurring along streams and moist areas in northern New Mexico, stinging nettle is a plant with medium to small toothed leaves covered with tiny hairs. When brushed against, the leaves and stems exude a stinging fluid that produces minor to moderate skin discomfort. The stinging sensation usually disappears fairly quickly. Keep a watchful eye around streams when hiking.

Stinging nettle (photo by Sally King)

Lack of Cell Phone Service

A number of areas covered by the guide are in areas that are out of cell phone range. Check your signal before setting out and at times while you are in the field to prevent the phone's battery from draining. We have noted the lack of specific coverage at individual sites; however, since coverage varies with provider, we may not have included an area where your specific phone does not provide coverage. Please check your own phone.

Flash Flooding, Debris Flow, Drought, and Wildfire

Much of New Mexico is considered to be a desert. However, the entire state recently has experienced a period of prolonged extreme drought. Fires have always been a part of nature's ecology, but recent drier conditions and insect infestations have contributed to increases in the frequency, intensity, and extent of wildfires. Spring and early summer wildfires yield to flash flooding during New Mexico's late-summer monsoon season. Locations downstream from burned areas are much more susceptible to flash flooding and debris flows. Flash floods can also occur where there is no obvious evidence of fire damage. Water running downhill through burned areas can create major erosion and pick up large amounts of ash, silt, rocks, and burned vegetation. The force of the rushing water and debris can damage or destroy culverts, bridges, roadways, and buildings even miles away from the burned area. Watch the sky for thunderstorms. A storm far upstream may flood an arroyo (dry wash) in seconds. Never park or camp in arroyos.

Monsoon Season

The Southwest regularly experiences what is known as the North American monsoon season. In New Mexico, the monsoonal pattern creates isolated, sometimes violent thunderstorms that usually form over mountains in the afternoon and then drift over the landscape before collapsing after sunset. The monsoon season usually begins in July and is over by early September. When traveling in the north-central New Mexico mountains in summer, watch the sky; mountain areas usually offer only one road down. Lightning poses a risk to hikers far from shelter. Flash flooding on slopes and in low areas can also be a threat. Check the media (broadcast, Internet, mobile phone apps) for weather forecasts.

Winter Driving

People visiting New Mexico for the first time are often surprised to see for-ested mountains with snowcapped peaks. Northern New Mexico's Sangre de Cristos are part of the Rocky Mountains. North-central New Mexico also has high deserts, vast plains, colorful sandstone canyons, and verdant river valleys with lakes formed by dams. This varied terrain can be a concern for winter driving. Many of the sites in this guide are on narrow, winding mountain roads. Please be cautious when driving on snowy or icy roads in winter. Check weather forecasts and the New Mexico Department of Trans-portation website (http://www.nmroads.com) or call its hotline (800–432–4269) before venturing into remote areas.

Altitude Acclimatization and Altitude Sickness

With one exception (Peña Blanca near Cochiti Lake at 5,236 feet), *all* loca-tions in this guide are more than one mile above sea level. The city of Santa Fe is at an elevation of 7,000 feet above sea level; Los Alamos, 7,320 feet; Taos, 6,967 feet; and even Las Vegas, on the Great Plains, is 6,420 feet above sea level. All these cities and towns are higher than Denver, Colorado. In ad-dition, the Santa Fe Ski Area, Taos Ski Valley, and Apache Creek are all above 10,000 feet. The hiking trails to see White-tailed Ptarmigan at Santa Barbara Ridge venture even higher at 12,000 feet.

After flying directly to New Mexico from sea level, it is recommended to take a day or two at around 7,000 feet to acclimatize to the altitude before attempting any very high-altitude birding such as at Santa Barbara Ridge (chapter 11). Before your trip to New Mexico, and while you are here, drink plenty of water. With New Mexico's low humidity and high altitude, you may need to drink twice as much water as at sea level. Carry water with you on hikes and in your vehicle. Water sources are limited at some of the bird-ing sites in this guide. Reducing exercise, alcohol, and caffeine intake is also recommended for the first few days.

The symptoms of altitude sickness or acute mountain sickness include throbbing headache, loss of appetite, nausea, vomiting, fatigue, insomnia, and dizziness. Seek medical attention for symptoms of severe altitude sick-ness, including confusion, blue lips and fingernails, and even pulmonary or cerebral edema.

Sun Protection, Clothing, and Shoes

Another effect of high altitude and arid climate is the lack of water vapor in the air. As a result, the sky may be bluer, but there is 25 percent less protection from the sun. Wear sunscreen even in winter. Consider a hat and long sleeves. Long pants and sturdy shoes are also recommended to defend against prickly desert plants and poison ivy.

Etiquette on Pueblos and Other Tribal Lands

Several of the sites in this guide are located on or adjacent to tribal lands (Ohkay Owingeh Fishing Lakes—Ohkay Owingeh Pueblo, chapter 3; and Cochiti Lake and Tetilla Peak Recreation Areas and Kasha-Katuwe Tent Rocks National Monument—Pueblo de Cochiti, chapter 8). Many Pueblos require a permit to photograph, sketch, or paint on location. Some Pueblos prohibit camera and cell phone use for photography or recording at all times. These photography restrictions are suspended at specific tribal birding areas described in this guide. Please check with the tribal office for the permitting process for cameras before entering different areas on Pueblo lands. The carrying or use of alcohol and drugs on Pueblos is strictly prohibited. Please be courteous and obey all posted signs when traveling to these sovereign tribal areas. Refer to the Indian Pueblo Cultural Center Web site (http://www.indianpueblo.org/) for more information.

Local Birding Information and Resources

Following are some of the important local resources (listed alphabetically) providing information for birders in the areas covered in this guide.

Army Corps of Engineers

The Army Corps of Engineers, in collaboration with the Wildlife Center, hosts a midwinter Eagle Watch at Abiquiu Lake each January.

Audubon New Mexico/Randall Davey Audubon Center and Sanctuary

The Center serves as the headquarters for Audubon New Mexico and its conservation and education activities around the state. Education is a primary goal of Audubon New Mexico and the Randall Davey Audubon Center and Sanctuary. The staff of environmental educators promotes awareness of the interrelationship between people, land, and wildlife through environmental education at its 135-acre wildlife sanctuary nestled in the foothills of the Sangre de Cristo Mountains and statewide. Except during January, volunteer naturalists lead bird walks each Saturday, on the grounds and into the Santa Fe Canyon Preserve. The area is one of Audubon New Mexico's 60 Important Bird Areas. Audubon New Mexico sponsors Christmas Bird Counts throughout the state. Its website provides information on birds in New Mexico (http://nm.audubon.org/birds-wildlife).

Bandelier National Monument

The Bandelier National Monument website (http://www.nps.gov/band/naturescience/summer-and-migratory-birds.htm) contains a wealth of information about the birds in the park, including a downloadable bird checklist, other fauna, plants, and the National Monuments ecosystems. The website also includes information that can be used by teachers. The Western National Parks Association operates the Trading Post, providing educational materials to park visitors to help them better understand the natural and cultural significance of the park. Its Facebook page announces special activities.

Bureau of Land Management

The Rio Grande Gorge Visitor Center is open seven days a week from 8:30 a.m. to 4:30 p.m., from Memorial Day to Labor Day. After Labor Day, the Center is open Friday through Sunday from 10:00 a.m. until 3:00 p.m. and has books, maps, and occasional programs. The Wild Rivers Visitor Center is open from Memorial Day through Labor Day, 9:00 a.m. until 6:00 p.m., and additional hours as staffing permits. After Labor Day, the Visitor Center is open as staffing permits and provides information about the geologic and natural history of the area.

Las Vegas National Wildlife Refuge

The Friends of Las Vegas National Wildlife Refuge sponsors programs and bird walks. Check its website (http://www.flvnwr.org/) to see what is happening.

New Mexico Ornithological Society (NMOS)

NMOS serves as the primary clearinghouse for information on the distribution, occurrence, and status of New Mexico birds. Information on rare-bird

sightings is summarized and published on its Web site twice a week (www.
nmbirds.org). Report unusual sightings to Matt J. Baumann, compiler, at
505–264–1052 or mb687@yahoo.com.

The publication *NMOS Field Notes* is printed quarterly for its members
and provides a seasonal overview of the changing patterns of New Mexico's
birdlife, including unusual records, breeding and wintering range changes,
and changes in seasonal occurrence and migration patterns. The *NMOS
Bulletin,* published four times a year, includes articles of scientific merit
concerning the distribution, abundance, status, behavior, and ecology of the
avifauna of New Mexico and its contiguous regions and news of interest to
the New Mexico ornithological community.

The organization also sponsors the New Mexico Bird Records Committee, which maintains the official list of New Mexico birds. The committee
requests documentation of all species on the Review List as well as New
Mexico records of species not currently on the official New Mexico State
List. A copy of the reporting form is found at the end of this chapter.

New Mexico State Parks

Several of the state parks provide a bird list. Check at the visitor center to
see if one is available.

Pajarito Environmental Education Center (PEEC)

PEEC is a focal point for information and programs related to environmental education in Los Alamos. Activities include PEEC Birders, a Los Alamos
bird list, and birding field trips. Visit its Web site for more information
(http://www.pajaritoeec.org).

Sangre de Cristo Audubon Society

Sangre de Cristo Audubon is a local chapter of the National Audubon Society. The organization has monthly meetings and sponsors field trips and
conservation efforts. For information on these activities, visit its Web site
(http://www.newmexicoaudubon.org/sdcas/).

US Forest Service

Sites in this book are located in both the Carson National Forest and the
Santa Fe National Forest. Their network of local Ranger Districts provides
maps, bird lists, and other information to make your visit more enjoyable.

New Mexico Rare/Unusual Bird Report

This form is intended as a guide in reporting observations of unusual birds. It may be used flexibly and need not be used at all. Leave blank any details not observed. Attach additional sheets if necessary. Attach any photographic, video recordings, or other documentation and send to Dr. Sartor O. Williams III, New Mexico Bird Records Committee, 1819 Meadowview Dr. NW, Albuquerque, NM 87104–2511, or sunbittern@earthlink.net.

Reporter (include middle initial, address, and email) _____

_____

Other observers who also identified bird _____

_____

Date(s)/times(s) when seen _____

Locality _____

Common or scientific name _____

Number/Age/Sex _____

Habitat _____

Conditions (weather, lighting) _____

Duration _____

Optical equipment (type, power, condition) _____

Distance to bird _____ Size and shape _____

Behavior (flying, feeding, resting) _____

Description of what was seen (total length, body bulk, bill, eye, leg characteristics, color, pattern of plumage) _____

_____

Was it photographed? ____ By whom? _____

Attached? _____

Voice _____ Previous experience with species _____

Basis for identification _____

Aids (book, other birders, illustrations) _____

_____

Positive of identification? _____ If not, explain _____

_____

Description written: from notes made during observation ___ notes made

after observation _____

Signature of reporter _____

Date/time written _____

This request for documentation should not be considered an affront, but an effort to substantiate a record by obtaining concrete evidence that may be permanently preserved for all to evaluate.

CHAPTER 3

Greater Santa Fe Area

General Overview

The seven sites in this chapter include two in or on the outskirts of the city of Santa Fe, four in the southern Sangre de Cristo Mountains east of Santa Fe, and one in the Española Valley north of Santa Fe. These sites encompass a wide range of habitats and vegetation zones.

La Cieneguilla Marsh, a wetland along the Santa Fe River, is surrounded by desert scrub habitat. Randall Davey Audubon Center and the Santa Fe Canyon Preserve are located in the piñon-juniper zone. Ohkay Owingeh Fishing Lakes are situated in a riparian area along the Rio Grande. As you travel along Hyde Park Road (NM 475) leading to the Santa Fe Ski Area (Ski Santa Fe), you will pass through ponderosa and mixed conifer habitats. The Santa Fe Ski Area, at an elevation of 10,350 feet, is located in the spruce-fir zone.

The temperature at the Santa Fe Ski Area may be as much as 40°F lower than at La Cieneguilla Marsh or Ohkay Owingeh Fishing Lakes. Be sure to check a weather forecast and road conditions before heading up NM 475 during winter months.

Randall Davey Audubon Center and Sanctuary and Santa Fe Canyon Preserve
Description
Just minutes up the canyon from the downtown plaza in Santa Fe, the Randall Davey Audubon Center and Sanctuary and the adjacent Santa Fe Canyon Preserve are wildlife treasures. The Center has been owned by the National Audubon Society since 1983 and is home base for Audubon New Mexico. The Preserve is owned and managed by The Nature Conservancy.

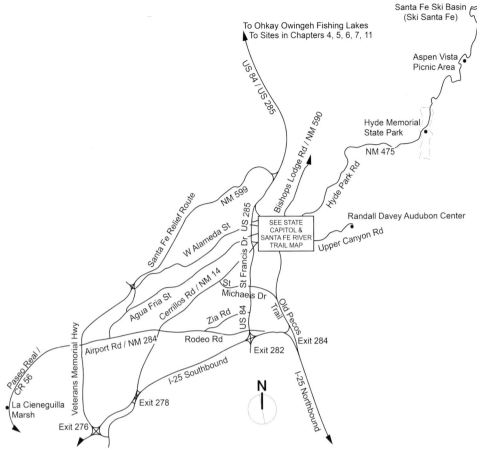

Map 2. Greater Santa Fe Area

The combined area of the Center and Preserve totals 660 acres and has been designated an Important Bird Area by National Audubon. This area contains a variety of habitats, including wetlands and ponds, montane riparian, desert scrub, piñon-juniper, and ponderosa and mixed conifer, each attracting associated bird species. Pick up a copy of the working bird checklist (more than 200 common and rare species) and the *Birding Santa Fe Canyon* brochure and trail map with species highlights. Volunteers lead bird walks every Saturday morning; check the Center website for starting times.

Randall Davey Audubon Center and Sanctuary: The Center encompasses 135 acres on the southwest corner of the Sangre de Cristo Mountains. Bounded by thousands of acres of Santa Fe National Forest and other protected lands, the Center includes the historic adobe Randall Davey home and studio (docent-led tours Fridays at 2:00 p.m.—fee), as well as lawns and gardens. Begin your visit by taking time to wander through the landscaped grounds, which include several old cottonwood trees, often enjoyed by several species of woodpeckers. Feeders and water features are strategically located year-round. When you reach the service road, follow it out to the main road. The brush piles and wet area to the left are favored by many species.

Meadow Loop and Bear Canyon Trails: From the Center's nature store, walk up the stone stairs to the piñon-juniper and meadow habitats above the Center. Bear left to reach "The Perch," a covered shelter providing a view of the upper Santa Fe River Canyon and Nichols Reservoir, part of the City of Santa Fe's municipal water system. This is a good location to scan for raptors soaring over the canyon. Beyond "The Perch," pick up Bear Canyon Trail, which drops down toward Bear Canyon Creek and then gradually climbs up the canyon through ponderosa pine and mixed conifer forests that attract montane species. Once the trail enters Santa Fe National Forest, it is unmaintained. Return by retracing your steps.

Santa Fe Canyon Preserve: Enter the Preserve from the trail located across from the paved driveway on the southwest side of the Randall Davey Audubon Center and Sanctuary. The steps lead first up and then down into the Preserve. The trail has a number of interpretive signs explaining the habitat types and history of the property. The trail crosses the streambed of the Santa Fe River (may be wet during spring runoff and summer rains) and a meadow and then forks at the base of the hill. Local birders recommend taking the left fork, which crosses the bypass channel of the Santa Fe River and enters a thick stand of willows before crossing the river again just above the 1881 Old Stone Dam. Listen and look for a variety of songbirds. There is an overlook at this point with benches providing good views into the dense cottonwood riparian area. This location can be particularly productive for warblers, especially during migration.

Follow the trail past the Old Stone Dam and gradually descend to a trail junction. At the junction, follow the "Enjoy the Mud" trail to the right down into the riparian area. The other trail is labeled "Avoid the Mud." The trail then follows the historic river channel in the bed of a former reservoir, passing a half dozen or more constantly evolving beaver ponds, magnets for birds. The two trails converge just before the trail rises up a short hill to another bench at the pond overlook. Listen for rails and search for waterfowl and cattail-loving birds. Just before dusk you may see a beaver in the pond. The trail crosses an area of four-wing saltbush, passes the former Two-Mile Reservoir spillway, and then leads to the Open Space Trailhead. As an alternative route, you can follow the Open Space Trail to Upper Canyon Road and turn left to return to the Center. As you walk along the road, stop at the intersection with Canyon Hill Road, where a stand of Siberian elms often attracts birds.

The Santa Fe Canyon Preserve is open daily year-round, dawn until dusk. The grounds of the Randall Davey Audubon Center are open 8:00 a.m. to 4:00 p.m. Monday through Saturday (closed Christmas and New Year's Day and the month of January). Check the Center's website for changes.

County: Santa Fe

eBird Hotspots: Randall Davey Audubon Center and Santa Fe Canyon Preserve

websites: http://nm.audubon.org/ and http://www.nature.org/ourinitiatives/regions/northamerica/unitedstates/newmexico/placesweprotect/santa-fe-canyon-preserve-1.xml

Target Birds

Black-chinned, Broad-tailed, and Rufous Hummingbirds Black-chinned and Broad-tailed Hummingbirds nest in the area. The Broad-tailed is the first to arrive, usually by mid-April, while the Black-chinned normally is not seen until late

Santa Fe Canyon Preserve

April or early May. Rufous Hummingbirds arrive in July after nesting in the Pacific Northwest. Calliope Hummingbirds may be seen on occasion during fall migration.

Steller's Jay and Western Scrub-Jay Both jays are common year-round. The Western Scrub-Jay is more prevalent in the piñon-juniper habitat, while Steller's Jay is more likely found along the Bear Canyon Trail.

Black-capped and Mountain Chickadees They can be found anywhere in the area and often visit the feeders on the Audubon Center grounds.

Spotted and Canyon Towhees Both species of towhee are year-round residents.

Townsend's Solitaire This bird can usually be found in fall and winter. As the birds arrive, they can be heard singing to establish their winter feeding territories.

Dark-eyed Junco During the winter, several junco subspecies can be observed, including Slate-colored (rare), Oregon, Pink-sided, and Gray-headed. The Gray-headed, which nests at higher elevations in the Sangre de Cristo Mountains, arrives first and is often present until the end of April or early May.

Black-headed Grosbeak This striking western grosbeak arrives at the end of April and is present through mid-September. Listen for its extended, robin-like song as it establishes and protects its breeding territory.

Other Birds

The area is blessed with many year-round species, including Mallard; Sharp-shinned, Cooper's, and Red-tailed Hawks; Eurasian Collared-Dove; White-winged Dove; Belted Kingfisher (absent in winter); Downy and Hairy Woodpeckers; Northern (Red-shafted) Flicker; Pinyon Jay (irregular); Black-billed Magpie; Clark's Nutcracker (irregular); American Crow; Common Raven; Juniper Titmouse; Red-breasted, White-breasted, and Pygmy Nuthatches; Brown Creeper; Bewick's Wren; Western Bluebird; American Robin; Curve-billed Thrasher; European Starling; Cedar Waxwing (irregular—follows berry crop); Song Sparrow; Red-winged Blackbird; Cassin's and House Finches; Red Crossbill (irregular); Pine Siskin; and Evening Grosbeak.

Species normally seen only in fall-winter-spring include Wild Turkey (not regular), Northern Pygmy-Owl (rare), Winter Wren (Santa Fe Canyon Preserve), Golden-crowned Kinglet, Hermit Thrush, White-crowned Sparrow, and American Goldfinch.

Birds that summer in the area include Turkey Vulture; Virginia Rail;

American Coot; Mourning Dove; Red-naped Sapsucker; Olive-sided, Cordilleran, and Ash-throated Flycatchers; Western Wood-Pewee; Say's Phoebe; Cassin's and Western Kingbirds; Plumbeous and Warbling Vireos; Northern Rough-winged, Violet-green, and Barn Swallows; House Wren; Blue-gray Gnatcatcher; Virginia's, MacGillivray's, Yellow, Grace's (Bear Canyon), and Black-throated Gray Warblers; Chipping Sparrow; Western Tanager; Brown-headed Cowbird; Bullock's Oriole; and Lesser Goldfinch.

During migration you might see Great Blue Heron; Osprey; Band-tailed Pigeon; Orange-crowned, Yellow-rumped, Townsend's (fall), and Wilson's Warblers; and Lincoln's Sparrow.

DIRECTIONS

In Santa Fe from Exit 282 on I-25, turn north on St. Francis Drive (US 84/285) and travel approximately 4 miles. Turn east (right, toward the mountains) on West Alameda Street, which becomes East Alameda Street. Alameda generally follows the Santa Fe River (a small stream) and a green space, Santa Fe River Park. After many four-way stops and traffic signals Alameda veers south and merges with Canyon Road. Very shortly, turn left onto Upper Canyon Road at the four-way stop. Follow Upper Canyon Road (eventually dirt) approximately 3 miles to its end at the Randall Davey Audubon Center parking lots.

PARKING

There are two lots at the Randall Davey Audubon Center (may be full with school groups and other special events). Parking for Santa Fe Canyon Preserve is also available at the Open Space Trailhead at the intersection of Upper Canyon Road and Cerro Gordo Road. Open Space Trailhead is signed as a high car-prowl area. Remove valuables and lock your car.

FEES

A donation is requested for visiting the Randall Davey Audubon Center grounds; trail admission donation for individuals and groups of fewer than eight people: $2.00 adults and $2.00 children; groups of eight or more: $5.00 adults and $1.00 children; free to Audubon members. There is no fee for visiting the Santa Fe Canyon Preserve.

SPECIAL CONSIDERATIONS AND HAZARDS

See chapter 2 for further safety guidelines.

- Black bears: Bears may be encountered along the river or seasonally at the Center's orchard.
- Cougars: There have been cougar sightings in the area.
- Trails: In both areas the trails can be muddy and difficult to navigate following rain or snowstorms and during spring runoff.
- Winter closures: The Randall Davey Audubon Center is closed during January. The road into both areas may be closed during snow.
- Habitat protection: To preserve sensitive habitat, please stay on formal trails.

FACILITIES

- Accessibility: There are level paths and benches on the main grounds of Randall Davey Audubon Center.
- Bear Canyon Trail: The path requires walking up and down steep stairs. There is a handrail only part of the way. Once up on the plateau, the trail is fairly level. The trail down into Bear Canyon is uneven and rocky at points.
- Santa Fe Canyon Preserve: The trail into the Preserve has a few shallow steps leading up and then several very steep stairs with no handrail heading down into the canyon.
- Restrooms: Randall Davey Visitor Center
- Water: There is a drinking fountain at Randall Davey Visitor Center and bottled water for sale in the nature store.
- Picnic tables: Randall Davey Audubon Center
- Pets: Pets are *not* allowed on the trails because bears and cougars may be present.

CAMPING

Camping is available in several Santa Fe National Forest campgrounds and in Hyde Memorial State Park along NM 475. There are also several private campgrounds in the Santa Fe vicinity.

GAS, FOOD, AND LODGING

There are numerous options for gas, food, and lodging in Santa Fe.

State Capitol and Santa Fe River Trail

Description

For visitors to Santa Fe who do not have the opportunity to travel outside the city, there are options for birding not far from the plaza and New Mexico State Capitol on foot or by public transportation. The plaza is the historical end point of the Santa Fe Trail, the trade route starting in St. Louis, Missouri, and traveled by settlers to New Mexico during the nineteenth century.

From the southeast corner of the plaza, walk south on Old Santa Fe Trail and connecting streets approximately 0.4 mile to the New Mexico State Capitol, referred to by New Mexicans as the "Roundhouse." This somewhat zigzag route is suggested for pedestrian travel only as several of the streets are one way in the other direction (see map). The Capitol Building entrance is located at the corner of Old Santa Fe Trail and Paseo de Peralta. Birding around the well-landscaped grounds can often be productive. If you have time, wander inside to see the vast collection of art by New Mexican artists.

After birding on the grounds, walk back toward the plaza until you reach the multiuse trail along the Santa Fe River at East Alameda Street. Turn

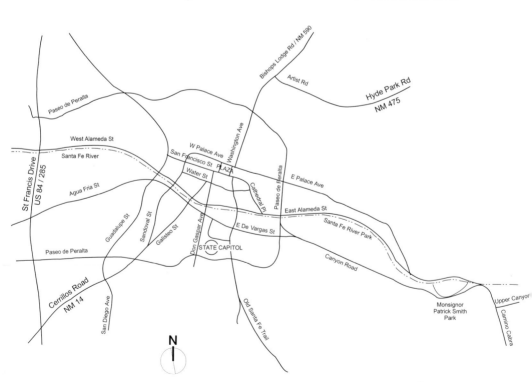

Map 3. State Capitol and Santa Fe River Trail

right (east) at the Santa Fe River Park and Trail. The trail/sidewalk continues 1.3 miles to Monsignor Patrick Smith Park. In addition to the riparian habitat along the river, the tall trees by the homes that border East Alameda Street on the north attract a variety of birds. Be careful to respect the privacy of the homeowners.

Another option is to take the Museum Hill bus (Route M) to Camino Cabra or drive east on East Alameda Street 1.3 miles to Monsignor Patrick Smith Park, where there is free parking. Bird around the park and then return along the river toward the plaza.

County: Santa Fe

eBird Hotspots: No

Target Birds

Northern (Red-shafted) Flicker Watch for the flash of red under the wings as it flies between trees. It can be found year-round.

Mountain Chickadee Often in a mixed flock, it can be found year-round.

Santa Fe River Trail

Cedar Waxwing Look for small flocks of these masked berry eaters in both deciduous and evergreen trees.

Spotted Towhee This year-round resident is most often seen scratching in the leaf litter under trees and shrubs.

Pine Siskin It is a prevalent year-round species in Santa Fe.

Lesser Goldfinch While it can be seen during the winter, it is more plentiful spring through fall.

Evening Grosbeak Although not predictable, it can be found year-round in Santa Fe.

Other Birds

During the winter, look for Williamson's Sapsucker (infrequent), Hairy Woodpecker, Common Raven, White-crowned and Song Sparrows, Ruby-crowned Kinglet, Western Bluebird, Hermit Thrush, Yellow-rumped Warbler, Dark-eyed Junco, and American Goldfinch.

During migration it is possible to see Western Wood-Pewee, Ash-throated Flycatcher, Violet-green Swallow, Chipping Sparrow, Western Tanager, and Black-headed Grosbeak.

In the summer, look for Turkey Vulture, Mourning Dove, Black-chinned and Broad-tailed Hummingbirds, and Barn Swallow.

Species that can be seen at any time of year include Cooper's Hawk, Rock Pigeon, Eurasian Collared-Dove, White-winged Dove, Downy Woodpecker, Western Scrub-Jay, Black-billed Magpie, American Crow, Black-capped Chickadee, Bushtit, European Starling, White-breasted Nuthatch, Bewick's Wren, Canyon Towhee, House Finch, and House Sparrow.

DIRECTIONS

On foot: The walking directions are described in the text.

By car: To reach Santa Fe Plaza from the intersection of I-25 (Exit 282) and St. Francis Drive (US 84/285), travel north 4 miles on St. Francis Drive. Turn right on West Alameda Street. Travel east approximately 0.5 mile to North Guadalupe Street. Turn left (north). Take the first right onto West San Francisco Street. The plaza will be on the left.

Public transportation: The Santa Fe bus system is called "Santa Fe Trails." The path of the Museum Hill/Route M bus travels along East Alameda Street to Camino Cabra before turning south and away from the birding areas described.

Several lots that charge a fee for parking are located near the plaza and Capitol. Free parking is available at Monsignor Patrick Smith Park.

FEES

None

SPECIAL CONSIDERATIONS AND HAZARDS

See chapter 2 for further safety guidelines.

- Vehicle and pedestrian traffic: Stay on paths and sidewalks. When you stop to look at a bird, be considerate of others who are using the paths and sidewalks.
- Private property: While it might be tempting to gaze into someone's yard to view a bird, please respect the person's privacy.

FACILITIES

- Accessibility: The walkways and paths are level and easy walking.
- Restrooms: Public restrooms are available in the Department of Tourism (491 Old Santa Fe Trail, across the street from the Capitol) in addition to a number of other public buildings. Their locations can be found at http://www.mellowvelo.com/maps/SantaFe_simple_city_map.pdf.
- Water: Available for purchase at businesses in the plaza and nearby areas
- Picnic tables: Several locations along the Santa Fe River Trail
- Pets: City ordinance requires that pets be kept on leash.

CAMPING

Closest camping is at Hyde Memorial State Park, 8 miles northeast of Santa Fe on NM 475.

GAS, FOOD, AND LODGING

Abundant options for gas, food, and lodging are available in Santa Fe.

La Cieneguilla Marsh

Description

This site is located in a marshy area along the Santa Fe River west of the city.

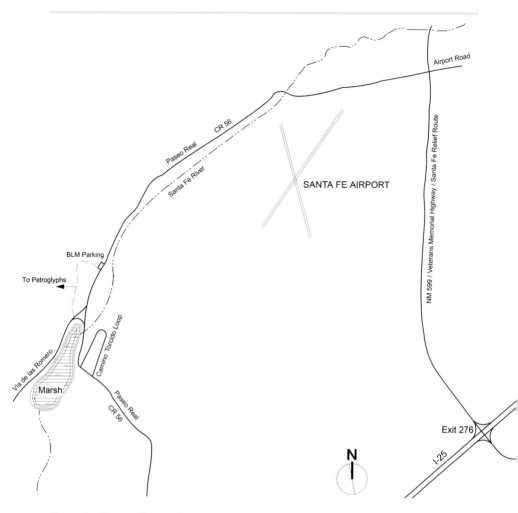

Map 4. La Cieneguilla Marsh

Directly behind the wetlands on land managed by the Bureau of Land Management (BLM) are basalt cliffs containing ancient Puebloan images (petroglyphs), including a number of birds, carved on the rocks. The riparian area along the river is bordered by upland desert scrub habitat. The river crosses under Paseo Real (County Road 56), which provides several viewing options.

Start your visit at the parking lot at the La Cieneguilla Petroglyph Site and follow the 0.42-mile trail through the desert scrub habitat to a dirt road, Via de Los Romero, which provides access to the northwest side of the wetlands. After exploring the northwest edge of the marsh, walk back around the north end on a curved cut-off dirt road leading back to Paseo Real, and walk southwest along the paved road. Stay alert for vehicular traffic as you bird along the highway, keeping well off the pavement on either side of Paseo Real.

If you drive past the marsh, you can turn around on Camino Torcido Loop just past the turn in the road. Heading back north, there are two small turnouts on the right (east). The first one is 0.1 mile past the curve in the road, and the other is 0.3 mile farther, just before the bridge over the Santa Fe River.

County: Santa Fe

eBird Hotspots: La Cieneguilla Marsh

Target Birds

Virginia Rail It is possible to find this secretive rail in the marsh vegetation year-round.

Cattails and Red-winged Blackbirds, La Cieneguilla Marsh

Marsh Wren This wren is often found in the cattails from October through April.

Common Yellowthroat It arrives in May and continues through mid-September.

Yellow-breasted Chat Look for it in the shrubs surrounding the pond. It is present from early May through mid-September.

Song Sparrow This winter resident begins arriving in September and is present through early May.

Red-winged Blackbird It is a year-round resident in the marsh. About an hour before sunset, large flocks will converge at the marsh for the evening.

Other Birds

In winter, look for Gadwall; Northern Shoveler; Green-winged Teal; Lesser Scaup; Northern Harrier; Sharp-shinned and Ferruginous Hawks; Northern Flicker; Merlin; Prairie Falcon; Belted Kingfisher; Ruby-crowned Kinglet; Dark-eyed Junco; and Lincoln's, Swamp, and White-crowned Sparrows.

Species during spring and fall migration include Swainson's Hawk; Peregrine Falcon; Orange-crowned, Virginia's, Yellow-rumped, and Wilson's Warblers; Green-tailed Towhee; and Chipping and Vesper Sparrows.

During summer it is possible to see Turkey Vulture (nests on the mesa behind the marsh); Common Nighthawk; Black-chinned Hummingbird; Northern Rough-winged, Barn, and Cliff Swallows; Northern Mockingbird; Yellow Warbler; Black-headed and Blue Grosbeaks; Lazuli Bunting; Brown-headed Cowbird; and Bullock's Oriole.

Year-round species include Mallard, Ring-necked Pheasant, Great Blue Heron, Cooper's and Red-tailed Hawks, American Coot, Eurasian Collared-Dove, White-winged and Mourning Doves, Greater Roadrunner, American Kestrel, Black and Say's Phoebes, Western Scrub-Jay, Common Raven, Bushtit, Bewick's Wren, American Robin, Curve-billed Thrasher, Spotted Towhee, Lesser Goldfinch, House Finch, and House Sparrow.

DIRECTIONS

From the intersection of St. Francis Drive (US 84/285) and I-25 (Exit 282) in Santa Fe, travel southwest on I-25 approximately 6 miles to Exit 276 (NM 599 / Santa Fe Relief Route / Veterans Memorial Highway). Turn right onto NM 599 and travel north approximately 3 miles. Turn left at Airport Road, which becomes Paseo Real (County Road 56). Follow CR 56 west and south 3–4 miles to the BLM Petroglyph Site parking lot on the right.

Parking is available in the BLM lot just north of the wetlands. There are also two small turnouts along Paseo Real: just past the bridge over the river and just before the road curves. Do not park on roads leading to private residences, including Via de Los Romero.

FEES
None

SPECIAL CONSIDERATIONS AND HAZARDS
See chapter 2 for further safety guidelines.

- Road traffic and shoulder hazards: Rural drivers may not be expecting pedestrians in this area. It is sensible to stay well off the pavement. In some areas, particularly on the bridge crossing the Santa Fe River, there are plants and roots in the pavement, making it difficult to stay on the shoulder.
- Snakes: Rattlesnakes are possible; keep your eyes and ears alert during warm weather.
- Ground squirrel / prairie dog burrows: There are numerous ground squirrel holes and partially filled abandoned holes in the area adjacent to the marsh and on the trail from the parking lot. Watch the ground when walking in this area.
- Harvester ants: Before stopping to observe a bird, look down to make sure you are not standing on a nest of these stinging ants.

FACILITIES

- Accessibility: The path is level along the road, although the shoulder is not very wide. The road to the rear of the marsh is level; however, it requires walking over livestock trails to reach the marsh.
- Restrooms: None on-site. The closest restroom is located at the mini-mart 0.2 mile east of NM 599.
- Water: None available
- Picnic tables: None available
- Pets: Not advised because of traffic

CAMPING
No camping site located nearby

Gas and food are available 0.2 mile east of NM 599. Numerous lodging opportunities are available along Cerrillos Road (NM 14) 4.5 miles east of Airport Road.

Along NM 475 (Hyde Park Road)

Description

Hyde Park Road (NM 475), designated the Santa Fe National Forest Scenic Byway, passes through four different life zones as it winds its way from downtown Santa Fe at an elevation of 7,000 feet to the ski area at 10,350 feet. There are a number of stops that provide excellent opportunities for viewing birds. Two stops have their own site descriptions: Aspen Vista Picnic Area and Santa Fe Ski Area. The initial part of the road leads through a residential area in the piñon-juniper-covered foothills before entering the Santa Fe National Forest. The following stops are recommended:

Dale Ball Sierra del Norte Trailhead: Located 2.7 miles along Hyde Park Road, this is a good trailhead to check for piñon-juniper species. Pinyon Jay and Clark's Nutcracker are often present in the fall.

Chamisa Trailhead and Trail (#183): Located just beyond mile marker 6, this is the first location within the national forest. The trailhead, at 7,835 feet, is in piñon-juniper habitat. From the trailhead a wide variety of piñon-juniper species can be seen, for example, Plumbeous Vireo, Cassin's Kingbird, and Black-throated Gray Warbler.

Little Tesuque Picnic Area and Black Canyon Campground: The picnic area will be on your left 1.5 miles beyond the Chamisa Trailhead (shortly past mile marker 8). The campground is 0.1 mile farther on the right side of the byway. There is a day-use parking area outside the entrance to the campground. Both stops are located at 8,200 feet in ponderosa habitat. Excellent birding is available at the picnic area. The Black Canyon Campground (open the last weekend in April to the last weekend in September) has a 1.5-mile trail (#181) that starts at the rear of the campground and loops through the nearby forest. Check the creek next to the parking area at the Black Canyon Campground in the spring for migrating passerines. From either location, you may expect several species, such as Red-naped Sapsucker, Cordilleran Flycatcher, Warbling Vireo, and Yellow-rumped (Audubon's) and Grace's Warblers.

Hyde Memorial State Park: Located 0.2 mile beyond Black Canyon, New Mexico's first state park offers an opportunity to view species found in ponderosa and mixed conifer habitat. There are several trails ranging from three easy walks (0.3 to 1.0 mile) to a strenuous 2.2-mile (one way) trail with a 1,000-foot elevation gain. Potential species include Cordilleran Flycatcher, Plumbeous and Warbling Vireos, Steller's Jay, all three nuthatches, Mountain Chickadee, Western Tanager, and Yellow-rumped and Orange-crowned Warblers.

Borrego (#150) / Bear Wallow Trailhead: The trailhead is on your left just past the state park RV camping area. While exploring the first mile of the trail, you will encounter many of the same birds mentioned at Hyde Park Memorial State Park. As you continue up Hyde Park Road, watch for Dusky Grouse beside the road, particularly in the early morning.

Big Tesuque Campground: Located at mile marker 13, this is a small but popular primitive campground located in mixed conifer habitat. Birds can sometimes be viewed along the creek that runs beside the northeast edge of the site.

US Forest Service Road 102 (FR 102): FR 102 intersects Hyde Park Road a half mile beyond Big Tesuque on the left. Park just beyond it in a turnout on the left and walk along the forest road for half a mile or so. Potentially

Big Tesuque Campground

any of the species that can be seen at the Aspen Vista Picnic Area (separate site description) may be seen here.

County: Santa Fe

eBird Hotspots: Santa Fe NF—Chamisa Trail, Santa Fe NF—Little Tesuque Picnic Area, and Hyde Memorial SP

Target Birds

Dusky Grouse This species is elusive. Most birders "stumble upon" it in the course of birding, hiking, or driving up Hyde Park Road, particularly in the early morning. The best opportunity for viewing it is in late April before nesting has begun or in August when the birds are traveling in family groups.

Flammulated Owl While it is occasionally reported at other locations, a good potential location is Hyde Memorial State Park from dusk into nightfall, mid-May through August.

Williamson's Sapsucker During the breeding season, it can be seen from the Black Canyon Campground / Little Tesuque Picnic Area up to the ski area. At other times of the year, look for it at the Chamisa Trailhead.

Red-naped Sapsucker Watch for it from late March through September from the Black Canyon Campground / Little Tesuque Picnic Area up to the Aspen Vista Picnic Area. It can be found at lower elevations during other times of the year.

Cordilleran Flycatcher It arrives in late May and can be found from the Black Canyon Campground / Little Tesuque Picnic Area up to the ski area.

Western Tanager It is a summer resident from the Black Canyon Campground / Little Tesuque Picnic Area up to the ski area. During migration it can be seen at lower elevations.

Other Birds

Species that can be seen during any time of year include Northern Goshawk; Red-tailed Hawk; Hairy Woodpecker; Pinyon (in piñon-juniper habitat) and Steller's Jays; Clark's Nutcracker; Common Raven; Mountain Chickadee; Red-breasted, White-breasted, and Pygmy Nuthatches; Brown Creeper; Cassin's Finch; and Pine Siskin.

During summer, look for Turkey Vulture; Cooper's Hawk; Broad-tailed Hummingbird; Northern Flicker; Western Wood-Pewee; Hammond's and Dusky Flycatchers; Plumbeous and Warbling Vireos; Violet-green Swallow;

Golden-crowned, and Ruby-crowned Kinglets; Western Bluebird; Townsend's Solitaire; Hermit Thrush; American Robin; Orange-crowned, MacGillivray's, Yellow-rumped, and Grace's Warblers; Dark-eyed Junco; and Black-headed and Evening Grosbeaks.

During migration it is possible to see Rufous and Calliope Hummingbirds and Townsend's (fall) and Wilson's Warblers.

DIRECTIONS

From the intersection of I-25 (Exit 282) and St. Francis Drive (US 84/285) in Santa Fe, travel north approximately 4.1 miles on St. Francis Drive to Paseo de Peralta (this is a loop road; take the "second" Paseo de Peralta off St. Francis Drive). Turn right and travel east 1 mile to Bishop's Lodge Road (NM 590). Turn left and travel north 0.2 mile. Turn right onto Artist Road, which immediately becomes Hyde Park Road (NM 475).

PARKING

Parking is available at the Chamisa Trailhead, Little Tesuque Picnic Area, outside the entrance to Black Canyon Campground, and near the visitor center at Hyde Memorial State Park.

FEES

Day use: Hyde Memorial State Park, $5.00

CAMPING

Black Canyon Campground and Hyde Memorial State Park

SPECIAL CONSIDERATIONS AND HAZARDS

See chapter 2 for further safety guidelines.

- Black bears: Although bears are possible at any time, during drought years conditions may have limited the bears' food supplies, motivating them to wander into more populated areas. Do not leave food unattended.
- Cougars: Cougars may be present, particularly in the early morning and in the evening.
- Poison ivy: Although not common, poison ivy can be found in moister areas.
- Seasonal fire restrictions: Restrictions may be posted, or the area

may be closed, if there is high fire danger. Check the Santa Fe National Forest website for alerts (http://www.fs.usda.gov/main/santafe/home).

- Winter: NM 475 is often snowpacked and icy during the winter months. It is advisable to check the New Mexico Department of Transportation website (www.nmroads.com) or call the hotline (800–432–4269) for current road conditions during winter months.

FACILITIES

- Accessibility: Birding at any of the trailheads or in the campgrounds is level.
- Restrooms: Little Tesuque Picnic Area, Black Canyon Campground day-use area, Hyde State Park, and Big Tesuque Campground
- Water: Black Canyon Campground, Hyde State Park, and Big Tesuque Campground
- Picnic tables: Chamisa Trailhead and Little Tesuque Picnic Area
- Pets: Santa Fe National Forest regulations require that pets be on leash.

CAMPING

Black Canyon Campground (reservations through website, www.recreation .gov), Hyde Memorial State Park, and Big Tesuque Campground

GAS, FOOD, AND LODGING

Abundant options for gas, food, and lodging are available in Santa Fe.

Aspen Vista Picnic Area
Description

At nearly 10,000 feet, with spruce-fir woodlands and stands of aspen, this site provides spectacular views across the mountain peaks and valleys. Take a moment to enjoy the view before starting your birding. To the right of the parking area is a well-maintained path that winds through the picnic area. There is an unmaintained trail leading southwest from this path that meanders through the woods, providing additional opportunities for viewing woodland birds.

To the left of the parking area beyond the gate is a US Forest Service access road that serves as the Aspen Vista Trail. The sign on the gate states "not for public use," which refers only to vehicle traffic. This popular trail, also used by mountain bikers and equestrians, particularly when the aspens are

View from Aspen Vista Picnic Area

at the height of fall color, gradually rises in elevation for 5.6 miles to Tesuque Peak at 12,040 feet. Walking even a mile or two allows for birding without being closed in by the vegetation.

County: Santa Fe

eBird Hotspots: Santa Fe NF—Aspen Picnic Area

Target Birds

Dusky Grouse While grouse can sometimes be seen in late April and early May, they are more easily seen in August and September when in family groups. This is not a guaranteed species; it is most likely to be seen in the early morning. Hiking several miles up to the tree line on Tesuque Peak may increase your chances of seeing Dusky Grouse.

Broad-tailed Hummingbird Arriving by mid-May, it nests here and can be seen zipping through the area until early September.

Warbling Vireo It arrives in early May and is present through mid-August.

Violet-green Swallow This high-elevation cavity nester can be seen from mid-May through the first week in August.

Yellow-rumped Warbler Listen for its warble high in the conifers starting in mid-May as it flits between the evergreen boughs. Although it is present through late August, it will be more difficult to locate once it stops singing in mid- to late July.

Green-tailed Towhee It prefers the brushy habitat below the parking area. Look for it between mid-May and mid-August.

Other Birds

Year-round birds include Hairy Woodpecker, Steller's Jay, Clark's Nutcracker (unpredictable), Common Raven, Mountain Chickadee, White-breasted and Red-breasted Nuthatches, Brown Creeper, Golden-crowned Kinglet, Pine Grosbeak (irregular), Cassin's Finch, Red Crossbill, Pine Siskin, and Evening Grosbeak (irregular).

Species present from mid-May through late summer include Band-tailed Pigeon, Red-naped Sapsucker, Downy Woodpecker, Northern Flicker, Western Wood-Pewee, Cordilleran Flycatcher, Pygmy Nuthatch, House Wren, Ruby-crowned Kinglet, Townsend's Solitaire, Hermit Thrush, American Robin, Chipping and Lincoln's Sparrows, Dark-eyed (Gray-headed) Junco, and Western Tanager.

DIRECTIONS

From the intersection of I-25 (Exit 282) and St. Francis Drive (US 84/285) in Santa Fe, travel north approximately 4.1 miles on St. Francis Drive to Paseo de Peralta (this is a loop road; take the "second" Paseo de Peralta off St. Francis Drive). Turn right and travel east 1 mile to Bishop's Lodge Road (NM 590). Turn left and travel north 0.2 mile. Turn right onto Artist Road, which immediately becomes Hyde Park Road (NM 475). Ascend NM 475 approximately 13 miles to Aspen Vista Picnic Area on the right.

PARKING

There is a large parking area.

FEES

None

SPECIAL CONSIDERATIONS AND HAZARDS

See chapter 2 for further safety guidelines.

- Altitude: Use precautions when hiking at this altitude.
- Black bears: Although bears are possible at any time, during drought years conditions may have limited the bears' food supplies, motivating them to wander into more populated areas. Do not leave food unattended.
- Cougars: Cougars may be present, particularly in the early morning and in the evening.
- Seasonal fire restrictions: Restrictions may be posted, or the area may be closed, if there is high fire danger. Check the Santa Fe National Forest website for alerts (http://www.fs.usda.gov/main/santafe/home).
- Winter: NM 475 is often snowpacked and icy during the winter months. It is advisable to check the New Mexico Department of Transportation website (www.nmroads.com) or call the hotline (800–432–4269) for current road conditions during winter months.

FACILITIES

- Accessibility: The trail through the picnic area is graded and fairly level. The trail beyond the picnic area is narrow and uneven and often requires stepping over logs. The Aspen Vista Trail is a maintained road.
- Restrooms: There is a wheelchair-accessible restroom in the picnic area.
- Water: None available
- Picnic tables: There are several along the path through the picnic area.
- Pets: Santa Fe National Forest rules require that pets be on leash.

CAMPING

Black Canyon Campground, Hyde Memorial State Park, and Big Tesuque Campground along Hyde Park Road

GAS, FOOD, AND LODGING

Abundant options for gas, food, and lodging are available in Santa Fe.

Santa Fe Ski Area (photo by Bonnie Long)

Santa Fe Ski Area
Description

Located at the end of the Santa Fe National Forest Scenic Byway, the national forest area around the Santa Fe Ski Area provides excellent high-altitude birding in spruce-fir habitat. The elevation at the parking lot is above 10,000 feet, and trails leading toward the Pecos Wilderness or to the top of the ski lift reach above 12,000 feet.

Start your visit by checking for birds around the perimeter of the parking area, as well as perusing the stream that flows through the wooded area between the upper and lower parking lots. Because many of the species nest in the uppermost branches of the conifers during late spring and early summer, this may necessitate familiarizing yourself with the songs of these breeding birds, as they may be difficult to locate. These include Golden-crowned and Ruby-crowned Kinglets and Orange-crowned and Yellow-rumped Warblers.

Starting in mid-May, inspect the structures around the base of the ski runs for Cordilleran Flycatcher. A service road leads from the back of the lodge to a maintenance building partway up the slope. During spring, summer, and fall, you can walk up the service road, listening and carefully scanning the trees on either side, and then return by walking down the trail under the chairlift. To reach the tree line, you can ride the chairlift during the height of the fall color season. (The dates are variable. Check the dates at http://skisantafe.com/index.php?page=fall-scenic-chairlift.) Species that can be seen at these elevations include Gray Jay and occasionally Pine Grosbeak.

The Winsor Trailhead (#254) is located at the lower end of the parking area next to the restroom. Cross over the wooden bridge, turn right, and walk along the wooded trail about 0.25 mile until you reach a meadow. At this point, the Winsor Trail becomes very steep with switchbacks. If you do not want to engage in strenuous hiking, take time to explore the meadow habitat, a good place for possible nesting Green-tailed Towhee and Chipping and Lincoln's Sparrows.

Sangre de Cristo Audubon installed feeders in 2013 near La Casa Day Lodge in hopes of attracting rosy-finches. They plan to maintain them for several years to determine if the finches will discover and utilize the feeders.

County: Santa Fe

eBird Hotspots: Santa Fe Ski Area

Target Birds

Dusky Grouse While it is a year-round species, it is most commonly encountered in April when females sometimes forage in open areas and in late summer when they search for food in clearings accompanied by their broods. Seeing Dusky Grouse is usually based on luck—being in the right place at the right time.

Cordilleran Flycatcher It nests at the ski area, often on and around structures. For the best opportunity to see one, look near the main buildings at the base of the ski area, as well as around the chairlift/restaurant buildings mid-slope.

Gray Jay While it is present year-round, Gray Jay is generally found at higher elevations near the top of the ski runs during most of the year. It frequently can be found near the parking lot during the winter months.

Clark's Nutcracker This jay is nomadic, depending on the availability of cones from large-seeded pines. Although it sometimes descends to lower elevations in the fall, it may be seen at this location at all times of year.

House Wren It nests in cavities at the ski area and can be seen foraging in low, scrubby habitat. It arrives in early May and is present through the end of August.

Golden and Ruby-crowned Kinglets Both species breed high in conifers from the base of the ski area to the top of the mountain. Ruby-crowned Kinglet arrives in late April to early May and begins migrating to lower elevations in September. While some Golden-crowned Kinglets are elevational migrants, they can be found at the ski area year-round.

Lincoln's Sparrow It nests in high mountain meadows. Look for it in the meadow near the parking lot / Winsor Trailhead, as well as along the edges of the meadow habitat under the ski lifts.

Red Crossbill Like the Clark's Nutcracker, this crossbill is dependent on the seed crop of its preferred pine species, making it nomadic in nature. Some years, large numbers of Red Crossbills can be found near the ski area. In other years there are relatively few.

Other Birds

Species that can be seen at any time of year include Northern Goshawk (irregular); Red-tailed Hawk; American Three-toed (irregular), Downy, and Hairy Woodpeckers; Steller's Jay; American Crow; Common Raven; Mountain Chickadee; Red-breasted and White-breasted Nuthatches; Brown Creeper; American Robin; Pine Grosbeak (irregular); Cassin's Finch; Pine Siskin; and Evening Grosbeak (irregular).

During fall migration it is possible to find Townsend's Solitaire and Wilson's Warbler.

From May through late summer, look for Turkey Vulture, Sharp-shinned Hawk, Band-tailed Pigeon, Broad-tailed Hummingbird, Williamson's and Red-naped Sapsuckers, Northern Flicker, Olive-sided Flycatcher, Western Wood-Pewee, Warbling Vireo, Violet-green Swallow, Townsend's Solitaire, Hermit Thrush, Orange-crowned and Yellow-rumped Warblers, Green-tailed Towhee, Chipping Sparrow, Dark-eyed Junco, Western Tanager, and House Finch.

DIRECTIONS

From the intersection of I-25 (Exit 282) and St. Francis Drive (US 84/285) in Santa Fe, travel north approximately 4.1 miles on St. Francis Drive to Paseo de Peralta (this is a loop road; take the "second" Paseo de Peralta off St. Francis Drive). Turn right and travel east 1 mile to Bishop's Lodge Road

(NM 590). Turn left and travel north 0.2 mile. Turn right onto Artist Road, which immediately becomes Hyde Park Road (NM 475). Follow Hyde Park Road approximately 15 miles to the end at the Santa Fe Ski Area.

PARKING

There is a large parking area.

FEES

None

SPECIAL CONSIDERATIONS AND HAZARDS

See chapter 2 for further safety guidelines.

- Altitude: Use precautions when hiking at this altitude.
- Black bears: Although bears are possible at any time, during drought years conditions may have limited the bears' food supplies, motivating them to wander into more populated areas. Do not leave food unattended.
- Cougars: Cougars may be present, particularly in the early morning and in the evening.
- Seasonal fire restrictions: Restrictions may be posted, or the area may be closed, if there is high fire danger. Check the Santa Fe National Forest website for updates (http://www.fs.usda.gov/main/santafe/home).
- Winter: NM 475 is often snowpacked and icy during the winter months. It is advisable to check the New Mexico Department of Transportation website (www.nmroads.com) or call the hotline (800–432–4269) for current road conditions during winter months.

FACILITIES

- Accessibility: While there is an elevation gain from the lower end of the paved parking area to the upper end, the surface is smooth. The service road is graded, and the elevation gain is gradual. Hiking along the Winsor Trail (#254) and under the chairlifts is more rigorous.
- Restrooms: Adjacent to the Winsor Trailhead
- Water: None, except for sale in La Casa Day Lodge when the ski area is open

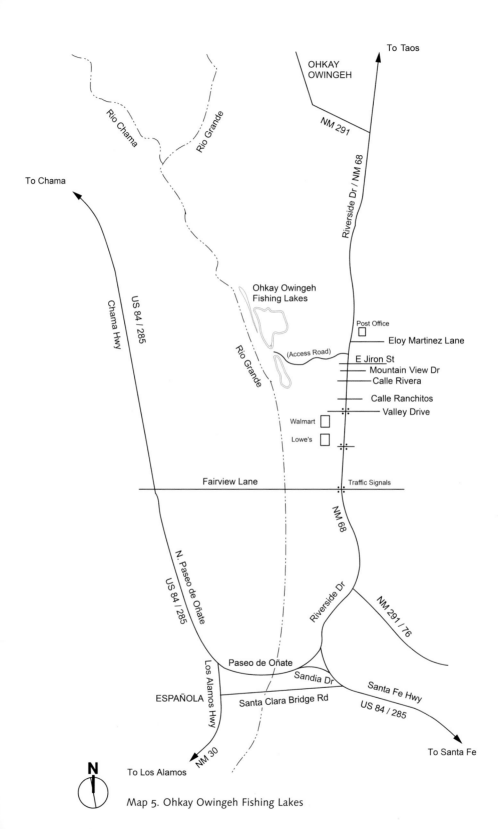

Map 5. Ohkay Owingeh Fishing Lakes

Ohkay Owingeh Fishing Lakes

- Picnic tables: Winsor Trailhead
- Pets: Santa Fe National Forest rules require that pets be on leash.

CAMPING

Black Canyon Campground, Hyde Memorial State Park, and Big Tesuque Campground along Hyde Park Road (NM 475)

GAS, FOOD, AND LODGING

Abundant options for gas, food, and lodging are available in Santa Fe, 15 miles below the ski area.

Ohkay Owingeh Fishing Lakes

Description

Known for wintering waterfowl, the Ohkay Owingeh Fishing Lakes provide excellent year-round birding just south of the confluence of the Rio Chama and the Rio Grande in the Española Valley at an elevation of 5,600 feet. The lakes, part of Ohkay Owingeh Pueblo and managed by Tsay Corporation, are stocked for fishing and attract a variety of species, including Double-crested Cormorant during spring and summer and waterfowl during the winter months.

In addition to scoping the two main lakes, check the riparian areas on both sides of the Rio Grande from the service road adjacent to the river. As you make your way north on the service road, you will come to an unmaintained 0.25-mile nature trail through a willow and cattail marsh, part of a long-term effort to restore the *bosque* (Spanish for "woodlands") along the river. This area is particularly productive in the early spring with migrating and nesting songbirds. By midsummer, the willows crowd the trail and make it more difficult to bird. After exploring the marsh, continue north on the service road to the boundary of the fishing lakes area. From this location, you have a good view north along the Rio Grande. Before leaving, walk along the cattail marsh leading from the entrance gate.

Summer hours are 7:00 a.m. to 6:00 p.m. on weekends and holidays, 8:00 a.m. to 5:00 p.m. on weekdays. If the gates are locked, visitors may park at the gate and walk in for birding. The entry fee may be assessed if a ranger arrives during your visit. During the winter months, the fishing lakes are open only on weekends from 9:30 a.m. to 6:30 p.m. and are closed when the lake surface is frozen. Because this is a popular destination during the height of fishing season, the best time to visit for birding is midweek.

County: Rio Arriba

eBird Hotspots: Ohkay Owingeh Fishing Lakes

Target Birds

Bufflehead **Normally not present until December, large numbers can be found in January and February, and a few remain through March. Look for them on the lakes as well as the river.**

Common Goldeneye **Large numbers are present during the winter months in the river and the lakes. Check the flocks carefully, as Barrow's Goldeneye has been found some winters.**

Common Merganser **It arrives in early November and can be found through early March.**

Double-crested Cormorant **One or two are present from March through October, although occasionally one will linger into December.**

Belted Kingfisher **It is found year-round, primarily along the river.**

Black-billed Magpie **The magpie is a year-round resident in the riparian areas.**

Blue Grosbeak **Look for it in the trees and shrubs, primarily along the river.**

Other Birds

In the winter, the fishing lakes and river should produce Gadwall, American Wigeon, Green-winged Teal, Canvasback, Redhead, Ring-necked Duck, Bufflehead, Hooded Merganser, Northern Harrier, Red-tailed Hawk, Common Raven, Marsh Wren, Western Bluebird, Townsend's Solitaire, American Robin, Yellow-rumped Warbler, Song and White-crowned Sparrows, Dark-eyed Junco, and American Goldfinch.

During migration, look for Cinnamon Teal; Ruddy Duck; Eared, Western, and sometimes Clark's Grebes; Osprey; Franklin's and Ring-billed Gulls; Northern Rough-winged, Tree, Violet-green, and Bank Swallows; and Ruby-crowned Kinglet.

Summer species include Snowy Egret, Green Heron, Black-crowned Night-Heron, Black-chinned Hummingbird, Western Wood-Pewee, Western and Eastern Kingbirds, Plumbeous Vireo, Barn Swallow, Common Yellowthroat, Yellow Warbler, Yellow-breasted Chat, Black-headed Grosbeak, and Lesser Goldfinch.

Year-round birds include Canada Goose, Mallard, Pied-billed Grebe, Great Blue Heron, Cooper's Hawk, American Kestrel, Greater Roadrunner, Northern Flicker, Black and Say's Phoebes, Western Scrub-Jay, Black-capped Chickadee, White-breasted Nuthatch, Spotted Towhee, Red-winged Blackbird, Great-tailed Grackle, and House Finch.

DIRECTIONS

From the intersection of St. Francis Drive (US 84/285) and Cerrillos Road (NM 14) in Santa Fe, travel north on US 84/285 approximately 25 miles to the city of Española. At the intersection with Paseo de Oñate, US 84/285 makes a left turn. Instead, continue north onto NM 68 (Riverside Drive). Follow NM 68, passing the Walmart on the left (traffic signal at Valley Drive). After approximately 2 miles beyond the intersection of US 84 and NM 68 (just past East Jiron Street), turn left and follow the unlabeled road west past the gate and cattail marsh to the fishing lakes entrance.

The accompanying map may be helpful. If you reach the post office on NM 68, you have gone past the access road to Ohkay Owingeh Fishing Lakes.

PARKING

Although there is no designated parking area, there is ample parking around both main lakes.

There is a $10.00 fee per car. Let the attendant know you are there to bird.

SPECIAL CONSIDERATIONS AND HAZARDS

See chapter 2 for further safety guidelines.

- Mosquitoes: Mosquitoes are prevalent along the trail in the willow and cattail marsh during the summer.
- Rattlesnakes: Rattlesnakes are possible.
- Bees and hornets: Be wary of stinging insects during the late summer and on the nature trail.
- Ants: There are frequently ants around the picnic areas during warm weather.
- Stay on roads and trails: Ohkay Owingeh Pueblo requests visitors not wander off the service road and trails to look for birds.
- Livestock and their droppings: While the Pueblo makes every effort to keep livestock off the property, they are often not contained on adjacent land.

FACILITIES

- Accessibility: The area is flat and level and easy for walking.
- Restrooms: Restrooms are available. If the area is closed, there are restrooms available at various businesses along Riverside Drive/NM 68.
- Water: None available
- Picnic tables: At scattered sites throughout the area
- Pets: Pets must be on leash, and owners must clean up waste.

CAMPING

Lakeside camping is available at unimproved sites around both lakes.

GAS, FOOD, AND LODGING

There are abundant opportunities for gas, food, and lodging in Española.

CHAPTER 4

Greater Taos Area

General Overview

The community of Taos, located on the Taos Plateau, can be reached from the Santa Fe area by one of two main routes: the "high" road that wanders through the foothills of the Sangre de Cristo Mountains, and the "low" road that follows the Rio Grande. The sites in this chapter are located along the low road (NM 68).

As you drive north from Española along NM 68, there are several turn-outs that might be productive for birding. The County Line Take-Out is 3.5 miles past the junction of NM 68 and NM 75 (which leads to the high road). This large, paved rest area has picnic tables and restrooms, is the terminus (or "take-out" point) for rafting trips along the Rio Grande, and provides an opportunity to bird in close proximity to the river and perhaps glimpse an American Dipper. There are six other turnouts before reaching the Rio Grande Gorge Visitor Center in Pilar.

Sites in this chapter provide examples of almost all habitats and vegetation zones. Three sites are located along the steep-sided Rio Grande Gorge, part of the Rio Grande del Norte National Monument, where riparian habitat lines the river, the walls of the basalt cliffs provide nesting habitat for swallows and raptors, and the mesas above the canyon provide habitat for species favoring Great Basin scrub vegetation.

Fred Baca Park within the town of Taos contains a 5-acre wetland and riparian area. The final site in this chapter is in and around the Village of Taos Ski Valley in the Sangre de Cristo Mountains.

The temperatures vary between Taos, the sites along the Rio Grande, and the Taos Ski Valley. While summer temperatures can reach the 80s in Taos, it cools down quickly once the sun goes down. The Taos area may have frequent

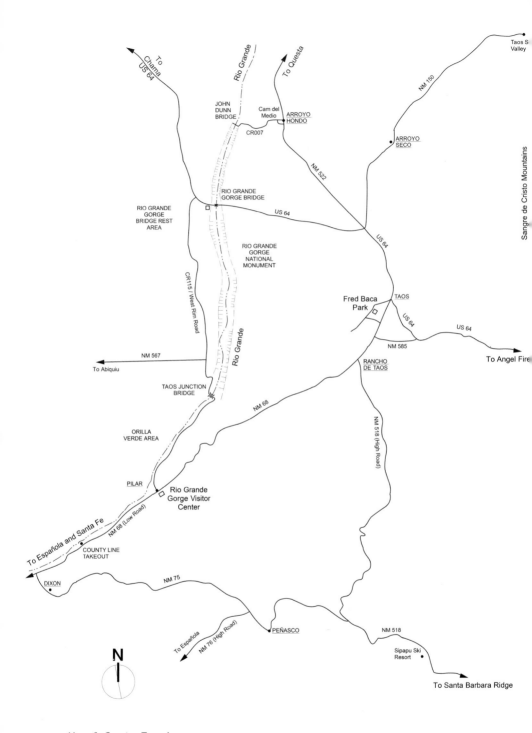

Map 6. Greater Taos Area

afternoon thundershowers in summer that seem to appear from nowhere and develop quickly. During the winter, daytime temperatures are often below freezing and night temperatures may dip below 0°F. It is advisable to check the New Mexico Department of Transportation website (www.nmroads.com) or call the hotline (800–432–4269) for current road conditions during winter months.

Elevations for these sites range from 6,100 feet along the river to 9,400 feet at the ski valley and 10,200 feet at Williams Lake Trailhead. See chapter 2 for information about high-altitude acclimatization.

Rio Grande del Norte National Monument: Orilla Verde Area
Description

The Orilla Verde Area is at the southern end of the Rio Grande del Norte National Monument and designated an Audubon New Mexico Important Bird Area. Before you turn onto NM 570 to drive through the canyon, stop at the Rio Grande Gorge Visitor Center, operated by the Bureau of Land Management, to pick up a map and a bird list. There is also information about the geology and natural history of the area.

As you enter the canyon, the noise of the traffic along NM 68 fades away. NM 570 winds through the canyon along the Rio Grande for 7 miles with both piñon-juniper and riparian habitats along the river and scrub habitat on the hillsides and tops of the mesas. Arrive early, as summer weekends can be busy with people camping, fishing, and rafting; however, during the week the canyon feels secluded.

The Rio Bravo Campground (1.7 miles) possibly provides the best area for riparian birding and good access to the river. Farther along, stop and explore the cottonwoods at the Petaca Campground. There are a number of turnouts along the river where you can stop and scan for American Dipper. After 5.9 miles you will arrive at a turnout that has particularly good habitat on both the river and hillside that attracts a variety of birds. At mile marker 6, the road forks at the Taos Junction Bridge.

Continue straight ahead and follow the Slide Trail (Old Highway 570) about 0.5 mile to the spot where the road is blocked and there is a parking area. Check for seasonal birds in the dense stand of trees along the Pueblo de Taos River before walking around the barrier. You can walk along the old road about 0.33 mile before the road is obliterated by the remains of a landslide. The trail navigates through the slide rubble and continues up the old

Orilla Verde Recreation Area

road all the way to the rim. Return and cross the bridge to the Taos Junction Campground, a good location for Canyon Wren.

You can return to NM 68 by retracing your route or continuing up the dirt road to NM 587 (West Rim Road, paved) that passes through Great Basin scrub habitat along the west mesa of the Rio Grande for 8.2 miles to US 64 near the Rio Grande Gorge Bridge Rest Area.

The West Rim Trail, heavily used by mountain bikers, extends north and south from the highway along the canyon's rim. Sagebrush Sparrows and Horned Larks can be found in the sage flats there.

The Rio Grande Gorge BLM Visitor Center at Pilar is open every day from 8:30 a.m. to 4:30 p.m., May 1–October 31, and 10:00 a.m. to 2:00 p.m., November–April. The recreation area day-use hours are from 6:00 a.m. to 10:00 p.m.

County: Taos

eBird Hotspots: Orilla Verde Recreation Area, Orilla Verde Recreation

Area—Rio Bravo Campground, and Orilla Verde Recreation Area—Taos
Junction Bridge
    website: http://www.blm.gov/nm/st/en/prog/recreation/taos/orilla_verde.
html

Target Birds
Common Goldeneye It is frequently seen along the river in January and
February.
Common Merganser It is seen along the Rio Grande from late September
through April.
Bald Eagle Often up to three Bald Eagles can be seen patrolling this stretch of
the river from December through February.
Ladder-backed Woodpecker It has become more prevalent in this location.
Look for it in scrub habitat.
Plumbeous Vireo It nests in piñon pine and juniper trees throughout the recre-
ation area.
Rock, Canyon, and Bewick's Wrens All three wrens are present year-round.
Canyon Towhee It is a year-round resident. Except when breeding males are
establishing their territory, Canyon Towhees normally are seen scurrying on
the ground. They often will rest under a parked vehicle.
Blue Grosbeak It breeds in the area, arriving in early May and staying through
late summer.
Bullock's Oriole An abundant breeder in riparian habitat, it is present from
early May through early September.

Other Birds
During the winter, look for Cackling Goose, Gadwall, American Wigeon,
Green-winged Teal, Ring-necked Duck, Bufflehead, Barrow's Goldeneye
(irregular), Pied-billed Grebe (irregular), American Coot, Mountain Chicka-
dee, Ruby-crowned Kinglet, Western and Mountain (West Rim) Bluebirds,
Townsend's Solitaire, White-crowned Sparrow, Dark-eyed Junco, and Pine
Siskin.
    The river serves as a migration corridor for Osprey; Rufous Humming-
bird (July through fall); Warbling Vireo; Orange-crowned, Yellow-rumped,
and Wilson's Warblers; Green-tailed Towhee; Western Tanager; and Lazuli
Bunting.
    Summer species include Turkey Vulture; Peregrine Falcon; Spotted Sand-

piper; White-throated Swift (Slide Trail); Black-chinned, Broad-tailed, and Rufous (July) Hummingbirds; Western Wood-Pewee; Black and Say's Phoebes; Ash-throated Flycatcher; Cassin's and Western Kingbirds; Northern Rough-winged, Violet-green, Barn, and Cliff Swallows; Blue-gray Gnatcatcher; Northern Mockingbird; Cedar Waxwing; Virginia's, Yellow, and Black-throated Gray Warblers; Yellow-breasted Chat; Chipping, Brewer's (West Rim), Lark, and Sagebrush Sparrows (West Rim); Black-headed Grosbeak; Brewer's Blackbird; Common and Great-tailed Grackles; Brown-headed Cowbird; and Lesser Goldfinch.

Birds that can be seen at any time of year include Canada Goose; Mallard; Scaled Quail; Great Blue Heron; Red-tailed, Sharp-shinned, and Cooper's Hawks; Golden Eagle; American Kestrel; Eurasian Collared-Dove (Pilar); Mourning Dove; Belted Kingfisher; Downy and Hairy Woodpeckers; Northern Flicker; Pinyon Jay; Western Scrub-Jay; Black-billed Magpie; American Crow; Common Raven; Horned Lark (West Rim); Black-capped Chickadee; Juniper Titmouse; Bushtit; White-breasted Nuthatch; American Dipper (irregular); Townsend's Solitaire; American Robin; European Starling (Pilar); Spotted Towhee; Song Sparrow; Red-winged Blackbird; House Finch; American Goldfinch; and Evening Grosbeak (irregular).

DIRECTIONS

From Santa Fe, at the intersection of St. Francis Drive (US 84/285) and Cerrillos Road (NM 14), travel north on US 84/285 approximately 25 miles to Española. At the intersection with Paseo de Oñate, US 84/285 makes a left turn. Instead, continue north onto NM 68 (Riverside Drive). Follow NM 68 approximately 30 miles to NM 570 at Pilar.

From Taos, at the intersection of US 64 and NM 68 (one block east of the plaza), travel south on NM 68 approximately 15 miles to NM 570 at Pilar.

PARKING

There is parking at day-use areas and at each campground.

FEES

Day-use fee or federal pass.

SPECIAL CONSIDERATIONS AND HAZARDS

See chapter 2 for further safety guidelines.

- Weather hazards: Lightning storms and flash floods occur in summer. The winding roads may be icy in winter. It is advisable to check the New Mexico Department of Transportation website (www.nmroads.com) or call the hotline (800–432–4269) for current road conditions during winter months.
- River currents: While the river through the recreation area is rated Class II (wide channel and easy rapids), during spring runoff wave swells may build up to 3 feet. Swimming or venturing into the river is not recommended.
- Rattlesnakes: Western diamondback rattlesnake is possible in the area. Be alert when hiking on the trails or exploring along the river.
- Poison ivy: It grows in spots along the river. Be particularly vigilant along the slide trail near the cement buffers and springs.
- Cell phone service: There is no cell phone coverage anywhere in the canyon.
- Cultural artifacts: The area has been inhabited for more than 10,000 years, and cultural artifacts continue to be found. The area is protected under the Archaeological Resources Protection Act, so removing artifacts is illegal.

FACILITIES

- Accessibility: While many of the campgrounds and day-use areas are level, some are rocky and uneven. Most of the trails lead over uneven terrain.
- Restrooms: Available at the Visitor Center and in each of the campgrounds
- Water: Available in most, but not all campgrounds, as well as at the Visitor Center
- Picnic tables: Available in all campgrounds and at the Visitor Center in Pilar
- Pets: Must be on leash and under control at all times

CAMPING

Seven campgrounds are located within the recreation area.

GAS, FOOD, AND LODGING

There is a small store with snacks in Pilar located across from the Visitor

Center on NM 68. Gas, food, and lodging are also available on NM 68 in Rancho de Taos, 12.2 miles north of Pilar; and gas and a mini-mart cafe are located in Velarde, approximately 15 miles south of Pilar.

Rio Grande del Norte National Monument: Rio Grande Gorge Bridge and Rest Stop

Description

These two birding areas are part of the Bureau of Land Management–managed Rio Grande del Norte National Monument and the Audubon New Mexico–designated Northern Rio Grande Important Bird Area. The Rio Grande Gorge is sometimes referred to as the "Grand Canyon of Taos." Visitors describe the view of the gorge from the bridge as awe inspiring, breath-

Rio Grande Gorge Bridge

taking, and amazing. The 650-foot depth of the gorge comes as a surprise as you drive west on US 64. The bridge, the second-longest truss bridge in the United States, is at the same level as the highway and is not apparent until you are almost on it—and you are unable to comprehend the view until you park and walk out onto the bridge. In addition to the inspiring geologic spectacle, there are several interesting specialty birds that reside in the canyon and on the plateau near the rest stop.

Park in the turnout on the northeast side of the bridge and walk along the path and out onto the bridge. For the more adventuresome, there are several waist-high "scenic porches" that jut out from the bridge. Listen for the descending trill of the Canyon Wren. Between April and late summer, White-throated Swifts and swallows can be heard chattering as they dart and swoop in the canyon.

After returning to your vehicle, drive across the bridge and pull into the rest stop on the other side. Walk the perimeter of the rest area, looking for Say's Phoebe that might be nesting near the buildings and inspect the grounds for sparrows. Then hike south along the Rim Trail a short distance for the opportunity to encounter Brewer's and Sagebrush Sparrows and Horned Lark.

County: Taos

eBird Hotspots: Rio Grande Gorge Bridge and Rio Grande Gorge Rest Stop

Target Birds

White-throated Swift It can be seen swooping through the canyon under the bridge from mid-April through August.

Say's Phoebe A grassland species, it might be found along the Rim Trail or at the rest stop, where it often nests on one of the buildings.

Violet-green Swallow More prevalent than the Cliff Swallow, also present at this location, it breeds in cliff cavities and is present between April and mid-September.

Canyon Wren Although it rarely sings during the winter, during the remainder of the year, its song can be heard from rock outcroppings in the gorge. It is more likely to be heard than seen.

Brewer's Sparrow This is one of the more reliable locations for this species. It breeds in the sagebrush habitat on the mesa on either side of the gorge and is present between mid-April and the end of August.

Lark Sparrow It is present from late April through the end of September. Look for it near the rest stop.

Sagebrush Sparrow This is a reliable location to find this species. Walk along the Rim Trail and look for it in the Great Basin shrub habitat on the mesa. When it briefly perches, it flicks its tail downward. Often it is seen scurrying along on the ground under the scrub. It arrives in late March and is present through late August/early September.

Other Birds

Birds found year-round at this location include Red-tailed Hawk, Northern Flicker, Western Scrub-Jay, Black-billed Magpie, American Crow, Common Raven, Horned Lark, Mountain Bluebird, and House Finch.

Species that can be seen during the summer include Turkey Vulture, Black-chinned Hummingbird, American Kestrel, Western Kingbird, Cliff Swallow, Rock Wren, Northern Mockingbird, Sage Thrasher (irregular), and Vesper Sparrow.

During migration, look for Rufous Hummingbird (fall), Yellow-rumped and Wilson's Warblers, Green-tailed Towhee, and White-crowned Sparrow.

DIRECTIONS

From central Taos, travel north 4 miles on Paseo del Pueblo Norte (US 64). Turn left at the traffic light, continuing on US 64. Travel 7.7 miles to Rio Grande Gorge Bridge.

PARKING

There is abundant parking at the rest stop. There is turnout parking on both sides of US 64 on the east end of the bridge. There is also a turnout on the north side of the west end of the bridge. Parking in these areas enables you to walk onto the bridge along a path that is protected from traffic.

FEES

None

SPECIAL CONSIDERATIONS AND HAZARDS

See chapter 2 for further safety guidelines.

- Harvester ants: You may encounter these hills of stinging ants on or adjacent to any of the trails.
- Precipitous cliffs: Take caution not to venture too close to the edge of the gorge while hiking along the Rim Trail.

- High winds: During the spring, the wind through the gorge may be strong enough to throw a person off balance. Use caution near the canyon rim or on the bridge road sidewalks.
- Acrophobia: Viewing the vast distance to the bottom of the gorge, combined with the trembling of the bridge as vehicles travel over it, can be disconcerting to those who have a fear of heights.
- Winter weather: It is advisable to check the New Mexico Department of Transportation's website (www.nmroads.com) or call the hotline (800–432–4269) before driving in the area during the winter.
- Rattlesnakes: Western diamondback rattlesnakes are possible in the area. Be alert for them in or near the rest area or when hiking on the Rim Trail.

FACILITIES

- Accessibility: Parking area, restrooms, and picnic tables are wheelchair accessible. Trails are rocky and uneven.
- Restrooms: Available at the rest stop
- Water: There is a drinking fountain at the rest stop.
- Picnic tables: Several are available at the rest stop.
- Pets: Due to the hazards, it is recommended to keep dogs on leash.

CAMPING

There are a number of RV parks in Taos. Public campgrounds are located in the Orilla Verde Area and along NM 585.

GAS, FOOD, AND LODGING

Private food-truck concessions may be present during the summer. Other food establishments, as well as gas and lodging, are available in Taos.

Rio Grande del Norte National Monument: Arroyo Hondo and John Dunn Bridge
Description
The area known as John Dunn Bridge, situated at the base of towering cliffs of layered volcanic and sedimentary rock, is a segment of a geologic rift that is part of the Rio Grande valley. Located at the north end of what is known to whitewater rafters as the Lower Taos Box, this area is part of the Rio Grande del Norte National Monument and designated by Audubon New Mexico as the Upper Rio Grande Gorge Important Bird Area. The cliffs

provide habitat and nesting areas for hawks and eagles, and the river is a migration corridor. The area is steeped in history and intrigue, as well as popular with anglers, rock climbers, rafters, and treasure-legend seekers.

Begin your birding as you turn west off NM 522 onto Old Highway 3 in the town of Arroyo Hondo. Seasonal waterfowl can be viewed at a pond and wetlands on private property on the north side of the road. Because this narrow road is busy and without a safe shoulder, to bird this area, park and bird from the post office parking lot or the lot at the market on NM 522. Do not encroach on private property. Just past the post office, Old Highway 3 then curves to the right and crosses the Arroyo Hondo. Turn left onto Cam del Medio (Lower Arroyo Hondo Road) and drive through the lower Arroyo Hondo valley. This dirt road travels past several small farms. While you may see birds in the fields and wetlands, the road lacks a sufficient shoulder lead-

Rio Grande at John Dunn Bridge

ing west to stop for birding. There are several turnouts on the left side of the road where it is convenient to stop on the return drive. After about 1.5 miles, the road turns to the right and merges with County Road B-007 (John Dunn Bridge Road). The farmhouses disappear, and the road continues into piñon-juniper habitat as it descends into the canyon. There are a few turnouts along the stretch of the road used by anglers where you can stop and scan the vegetation along the Arroyo Hondo and adjacent cliffs as you work your way toward the Rio Grande and John Dunn Bridge.

During warm weather, it is recommended that you arrive early in the morning, as the Rio Grande area can be crowded with anglers, rock climbers, or kayakers. It is also the put-in spot for whitewater float trips into the Lower Taos Box. There are areas to park on the left just before the bridge or on the right after crossing the bridge. You can explore the areas along the river, including sandbars, using the fishing trails.

County: Taos

eBird Hotspots: John Dunn Bridge and trails, Arroyo Hondo—Lower Arroyo Hondo Road, and Arroyo Hondo—pond and wetlands

Target Birds

Golden Eagle Golden Eagle nests in the cliffs up and down the gorge and often can be seen from this location.

Rock and Canyon Wrens Canyon Wren, found year-round in the canyon areas, often is first identified by its descending call. Rock Wren is often absent during the winter months, but one or two linger during some winters.

American Dipper It is possible to see one along the Arroyo Hondo, as well as the Rio Grande. It is often easiest to spot one during nesting when dippers make frequent trips to and from the nest site. This area provides good nest sites—under bridges and crevices on cliffs adjacent to a roaring river or stream. Nesting may begin as early as late March or as late as early May, depending on weather and water conditions. Dippers occasionally produce second broods.

Yellow Warbler This colorful warbler is a summer resident and breeder both in the area of the John Dunn Bridge and along the route to the bridge. Look for it at midlevel in the willows.

Song Sparrow It is a year-round species in wetlands and in brushy areas along the Arroyo Hondo and Rio Grande.

Other Birds

During winter, look for Green-winged Teal and Common Merganser on the Rio Grande. Townsend's Solitaire and Ruby-crowned Kinglet can be seen between October and March/April.

Species seen during the summer include Green Heron, Turkey Vulture, White-throated Swift, Broad-tailed Hummingbird, Western Wood-Pewee, Cordilleran Flycatcher, Northern Rough-winged and Violet-green Swallows, Yellow-breasted Chat, Black-headed Grosbeak, and Brown-headed Cowbird.

During migration you might see Spotted Sandpiper, Wilson's Snipe, House Wren, MacGillivray's and Wilson's Warblers, and Western Tanager.

Year-round species include Mallard, Great Blue Heron, Red-tailed Hawk, Black and Say's Phoebes, American Robin, and Red-winged Blackbird.

DIRECTIONS

From Taos, travel 4.1 miles north on US 64W to the intersection with NM 522 and NM 150, where US 64 turns west toward the Rio Grande Gorge Bridge. Do not turn; drive straight at the traffic signal on NM 522 to the small town of Arroyo Hondo, approximately 6.8 miles.

PARKING

The post office parking lot in Arroyo Hondo or the lot at the market on NM 522 can be used to bird the pond and wetlands. There is informal parking on both sides of the John Dunn Bridge.

FEES

None

SPECIAL CONSIDERATIONS AND HAZARDS

See chapter 2 for further safety guidelines.

- Gnats and flies: In warm weather, there can be swarms of distracting flying insects.
- Road conditions: The dirt road leading to the bridge may be difficult to drive during and after heavy rains or in the winter. Rock slides in the canyon may close the road.
- Cell phone service: There is no cell phone coverage in the canyon.
- Rattlesnakes: Be alert for rattlesnakes, which are possible along the trails.

- Accessibility: The fishing trails are uneven and often difficult to traverse.
- Restrooms: There is a restroom in the west parking area.
- Water: None available
- Picnic tables: None available
- Pets: Due to the potential for snakes, pets should be kept on leash.

CAMPING

There is no camping available on-site.

GAS, FOOD, AND LODGING

Nearest gas, food, and lodging are in Taos.

Fred Baca Park

Description

Located on the southwestern edge of town not too far from the Taos Plaza, Fred Baca Park has a 5-acre wetlands area with a boardwalk stretching across its length that hosts a variety of birds at all times of the year. Each season has its own signature and specialty birds. The deciduous trees planted around the park attract a variety of riparian species. Because this park is often busy with the activities associated with a city recreation area, the best time to visit is in the early morning or during the week. The park is open from 8:00 a.m. until 8:00 p.m. during the summer and from 8:00 a.m. until 5:00 p.m. in winter.

When you arrive, take time to bird the trees that line the perimeter of the parking area for woodpeckers and Black-billed Magpie before walking across the path to the wetlands.

The shrubs and willows that serve as a barrier between the park and wetlands are also worth checking before turning onto the boardwalk. As you step onto the plank path, you will be greeted almost immediately by seasonal birdcalls and songs. Walk slowly as you listen and watch for movement. After meandering the length of the park on the boardwalk that ends at a viewing area overlooking cattails and a beaver pond, return the way you came. When you get to the spot where the boardwalk turns left to return to the park, turn right and step off the boardwalk and walk a short distance on the trail that leads behind the wetlands. This is another area that can be productive for woodpeckers and other cavity nesters.

Fred Baca Park

After birding the park, walk to the entrance drive, cross Camino del Medio, and then walk carefully along the shoulder to a marshy area downstream from Fred Baca Park where Camino del Medio intersects the Rio de Fernando de Taos. You can walk across the bridge over the Rio de Fernando de Taos to a narrow turnout where you can safely bird. Willow Flycatcher nests here. Do not intrude on private property along the road.

County: Taos

eBird Hotspots: Fred Baca Park

Target Birds

Virginia Rail The cattail marsh at the end of the boardwalk is a good place to listen for, and hopefully see, Virginia Rail. It has been recorded at this location between February and November.

Sora It also inhabits the cattail marsh and is present during the summer months.

Lewis's Woodpecker While not guaranteed, it often can be seen in the deciduous trees bordering the parking lot or at the back of the marsh.

Willow Flycatcher Both migrants and breeding birds (June 15–July 20)—likely federally endangered "Southwestern" subspecies—are possible in the marsh southwest of Camino del Medio along the Rio de Fernando de Taos. It is illegal to use playback to call it in.

Black-billed Magpie This imposing corvid is found year-round in the park.

Gray Catbird Look deep in the understory, where it lurks in the wetlands from early May through mid-August.

Evening Grosbeak This roving species that travels in flocks might be seen at any time of year.

Other Birds

During the winter months, look for Green-winged Teal, Northern Harrier, Marsh Wren, Townsend's Solitaire, Hermit Thrush, White-crowned Sparrow, and Dark-eyed Junco.

Migration brings White-faced Ibis (spring); Killdeer; Band-tailed Pigeon; Broad-tailed and Rufous Hummingbirds (mid-July through fall); Warbling Vireo; Tree and Cliff Swallows (fall); Blue-gray Gnatcatcher; Orange-crowned (spring), MacGillivray's, Yellow-rumped, and Wilson's Warblers; Chipping and Brewer's Sparrows; Western Meadowlark; Yellow-headed and Brewer's (fall) Blackbirds; and Cassin's Finch.

Species seen during the summer include Cinnamon Teal, Green Heron, Sharp-shinned and Swainson's Hawks, American Kestrel, Mourning Dove, Black-chinned Hummingbird, Red-naped Sapsucker, Western Wood-Pewee, Black Phoebe, Cassin's and Western Kingbirds, Violet-green and Barn Swallows, Virginia's and Yellow Warblers, Common Yellowthroat, Yellow-breasted Chat, Spotted Towhee, Lark Sparrow, Common Grackle, Brown-headed Cowbird, and Bullock's Oriole.

A variety of birds are present year-round, including Mallard, Great Blue Heron, Red-tailed Hawk, Rock Pigeon, Eurasian Collared-Dove, Belted Kingfisher, Downy Woodpecker, Northern Flicker, American Crow, Common Raven, Black-capped Chickadee, American Robin, European Starling, Cedar Waxwing (irregular), Song Sparrow, Red-winged Blackbird, Great-tailed Grackle, House Finch, Pine Siskin, and American Goldfinch.

DIRECTIONS

From Taos Plaza, travel west on Ranchitos Road (NM 240) approximately 0.7 mile to La Posta Road. Turn left and travel south, bearing right immediately onto Camino del Medio. Travel southwest approximately 0.5 mile to Fred Baca Park entrance on the left.

PARKING

There is ample paved parking.

FEES

None

SPECIAL CONSIDERATIONS AND HAZARDS

See chapter 2 for further safety guidelines.

- Mosquitoes: There may be mosquitoes in the summer along the boardwalk or near the marsh on Camino del Medio.
- Winter weather conditions: The boardwalk may be icy or closed. The material used to construct the boardwalk can be slippery. It is advisable to check the New Mexico Department of Transportation website (www.nmroads.com) or call the hotline (800–432–4269) for current road conditions during winter months.
- Traffic along Camino del Medio: If you bird the marsh along the road, stay well off the highway. The road curves, and vehicles often travel fast here.

FACILITIES

- Accessibility: There are paved trails in selected areas. Although the boardwalk is fairly level, it is prone to being slippery and can be a hazard when wet or icy.
- Restrooms: There is an accessible restroom near the parking area.
- Water: None available (If you bring your own, note that glass containers are prohibited.)
- Picnic tables: There are several picnic tables and a shelter.
- Pets: Park rules require that all dogs be on leash.

Taos Ski Valley

CAMPING

There are a number of RV parks in Taos. Public campgrounds are located in Orilla Verde Area and along NM 585.

GAS, FOOD, AND LODGING

There are ample opportunities for gas, food, and lodging within Taos.

Taos Ski Valley

Description

From Taos, travel north on US 64 to the intersection with NM 150 (Taos Ski Valley Road). Turn right onto this gently curved highway through four life zones that gradually transports you from piñon-juniper to spruce-fir habitat at the Village of Taos Ski Valley at 9,325 feet. In the nineteenth century, the area occupied by Taos Ski Valley was once a mining town, named Twining after its founder. Birding along the trail to Williams Lake can take you to the beginning of alpine tundra habitat. An online map of the Taos Ski Valley roads and trailheads is available at the village's website (http://www .taosskivalley.com/uploads/gfx/VTSV_VillageMap_p2.pdf).

During the spring, summer, and fall, start your birding in the village. Park in one of the spacious lots and then walk toward the guard shack on Sutton Place. The beginning of the J. R. Ramming Memorial Nature Trail will be on your right before Sutton Place crosses the Rio Hondo. The well-planned and maintained nature trail follows the river and hosts Hermit Thrush, Orange-crowned Warbler, Western Tanager, and other high-elevation riparian species.

Next, locate Twining Road at the northeast corner of the parking area. Follow the signs to the Bavarian Lodge and to the Williams Lake Trailhead. Drive 0.9 mile along Twining Road to Zap's Road and continue another 0.9 mile (Zap's Road becomes Porcupine Road). Turn right at Kachina Road; continue a short distance to the Williams Lake Trailhead parking area. Take a few minutes to bird the perimeter before starting to walk down the trail/service road. Stop to bird the wetlands area at the base of the hill. A trail leads around a pond where you might see Lincoln's and White-crowned Sparrows. Continue along the trail/service road toward the Bavarian Lodge and upper ski lifts.

The trail (#62) to Williams Lake starts near Kachina Chairlift #4 at an altitude of 10,200 feet (2.5 miles from parking area). The trail is not obvious; look for a small round sign with a hiker symbol and an arrow below it on the side of the chairlift building. The trail starts climbing and follows a service road along East Fork Creek 0.25 mile, where you will see the Williams Lake Trail sign. The trail continues climbing through stands of spruce and fir, enters the Wheeler Peak Wilderness Area, and finally opens into a meadow before dropping down to the lake at an altitude of 11,040 feet. The trail is very popular, especially on summer weekends, which may make it difficult for birding.

On your return to the Village of Taos Ski Valley, you can stop at Beaver Pond (0.9 mile from parking area). The stream leading to and from the pond is a good place to look for American Dipper.

An alternative for viewing high-altitude birds is to ride the chairlift— either round-trip or one way and hike down. The chairlift is open Thursdays through Mondays from the end of June through Labor Day.

If you visit during the winter, one of the main attractions is the three

Gray Jay (photo by Jim Tuomey)

species of rosy-finch that frequent the feeders at the Kandahar Condominium office, a "ski-in-ski-out" winter lodging that sits on the hillside overlooking the Children's Ski Center. Parking during the winter can be challenging; however, shuttles transport visitors from the far-flung parking areas to the ski valley village center.

County: Taos

eBird Hotspots: Taos Ski Valley, Taos Ski Valley—Kandahar Condominiums, and Williams Lake

Target Birds

Gray Jay It is a year-round bird in the ski valley. Likely locations are from the hiker's parking lot to the Williams Lake Trailhead and along the forested parts of the trail.

Clark's Nutcracker It is also found year-round in the same locations as the Gray Jay.

American Pipit It nests in subalpine meadows and alpine tundra, near and at higher elevations than Williams Lake. It often can be seen near Williams Lake in late summer.

Orange-crowned Warbler A good place to see this understory warbler is along the J. R. Ramming Memorial Nature Trail between mid-May and mid-September.

Green-tailed Towhee Look for it near forest edges in the ski valley and all the way up to Williams Lake between June and September.

Lincoln's Sparrow It nests in wet, boggy areas. Look for it in the meadow near Kachina Road and at the base of the upper ski lifts.

Gray-crowned, Black, and Brown-capped Rosy-Finches All three species of rosy-finch come into the feeders maintained by the Kandahar Condominiums during the winter months. They arrive by late November and can be present through late February/early March.

Other Birds

During the summer you can expect to see Turkey Vulture, White-throated Swift, Broad-tailed Hummingbird, Western Wood-Pewee (along nature trail), Cordilleran Flycatcher, Violet-green Swallow, House Wren, Ruby-crowned Kinglet, Townsend's Solitaire, Hermit Thrush, American Robin, Wilson's Warbler, White-crowned Sparrow, Dark-eyed Junco, Western Tanager, and Pine Siskin.

During migration, look for Townsend's Warbler (fall). In September you might also see young Golden Eagles and Cooper's Hawks that have dispersed from their nesting sites and are exploring the area. Golden Eagle is most likely to be seen near Williams Lake, where it hunts marmots.

Year-round species include Dusky Grouse (although normally seen only in late summer), Red-tailed Hawk, Hairy and American Three-toed Woodpeckers, American Crow, Common Raven, Mountain Chickadee, White and Red-breasted Nuthatches, American Dipper, Pine and Evening (irregular) Grosbeaks, and Red Crossbill.

DIRECTIONS

From the intersection of US 64 and NM 68 just east of the plaza in the town of Taos, travel north 4.1 miles on Paseo del Pueblo Norte (US 64W). At the traffic signal, turn right on NM 150 and ascend the canyon 14.6 miles to the Village of Taos Ski Valley.

PARKING

There is a large parking area near the village and a medium-sized dirt lot at Carson National Forest—Williams Lake Trailhead (can be crowded if you arrive midday).

There is a fee to ride the chairlift in summer (and winter). Parking at Taos Ski Valley is free year-round.

SPECIAL CONSIDERATIONS AND HAZARDS
See chapter 2 for further safety guidelines.

■ Altitude: Altitude sickness can be a hazard for individuals not used to high elevations (elevation 9,300 feet at ski valley; 10,200 feet at Williams Lake Trailhead; 11,161 feet at the highest point on Williams Lake Trail).

■ Winter weather: The roads can be snowpacked and icy. It is advisable to check the New Mexico Department of Transportation website (www.nmroads.com) or call the hotline (800–432–4269) for current road conditions in winter.

■ Avalanche hazard: Check with the ski patrol before hiking/snowshoeing in winter.

■ Thunderstorms: Thunderstorms can materialize out of a blue sky during the afternoon in the summer.

■ Cell phone service: While there is coverage near the main chairlifts, by the time you reach the Williams Lake Trailhead, there is no cell service.

■ Environmental impact: Please stay on designated trails to prevent erosion and damage to native plants.

■ Bears: Black bears are possible. Do not leave food unattended.

■ Cougars: The US Forest Service sign at the Williams Lake Trailhead parking lot alerts hikers to the possibility of cougars, particularly if you are hiking at dawn or dusk.

FACILITIES

■ Accessibility: The nature trail along the creek in the village is compacted gravel and even, with places to sit along the trail. The trail/service road from the Williams Lake Trailhead down to the Bavarian Lodge and upper ski lifts is even and has a relatively gentle grade. The trail to Williams Lake is not difficult; however, it is rocky and uneven.

■ Restrooms: There are restrooms in the village near the main ski lifts, at Twining Campground, and next to the Williams Lake Trailhead parking area.

- Water: Water can be purchased in the village.
- Picnic tables: There are picnic tables in the village near the rope tow lifts, at Twining Campground, and next to the Williams Lake Trailhead parking area.
- Pets: Carson National Forest rules require that dogs be leashed when on all trails. Please pick up and pack out pet waste.

CAMPING

There is a small primitive campground on Twining Road near the Bull-of-the-Woods Trailhead. There are three other small campgrounds along NM 150.

GAS, FOOD, AND LODGING

The closest gas station is in El Prado, just north of Taos on US 64. Food and lodging are available in Taos Ski Valley, Arroyo Seco, and Taos.

CHAPTER 5

Enchanted Circle Scenic Byway and Nearby Areas

General Overview

The Enchanted Circle Scenic Byway makes a wide circle around Wheeler Peak, New Mexico's highest mountain. The six site descriptions in this chapter start north of Taos near Questa, then follow NM 38 through the Carson National Forest and the village of Red River in the Sangre de Cristo Mountains, to the village of Eagle Nest and Angel Fire in the Moreno Valley, returning to Taos through Taos Canyon, which is primarily bounded by private property. We have included Cimarron Canyon State Park east of Eagle Nest, not located directly along the Enchanted Circle. Other areas also on the Enchanted Circle Scenic Byway (Taos area sites and Arroyo Hondo) are included in chapter 4.

These sites encompass a wide variety of habitats: Great Basin scrub, piñon-juniper, ponderosa pine, mixed conifer, and spruce-fir.

Weather conditions vary by individual site during spring, summer, and fall and are addressed in the individual descriptions. Winter weather can be severe. Visitors should check road conditions (New Mexico Department of Transportation website, www.nmroads.com; or call the hotline, 800–432–4269) before driving to any of these sites during the winter.

Red River State Fish Hatchery
Description
As you leave the piñon-juniper woodlands along NM 522, ponderosa pines become more prevalent 20.5 miles north of Taos as NM 515 heads down toward the hatchery. The New Mexico Game and Fish Department manages the fish hatchery, which is located in a red-rock canyon along the Red River

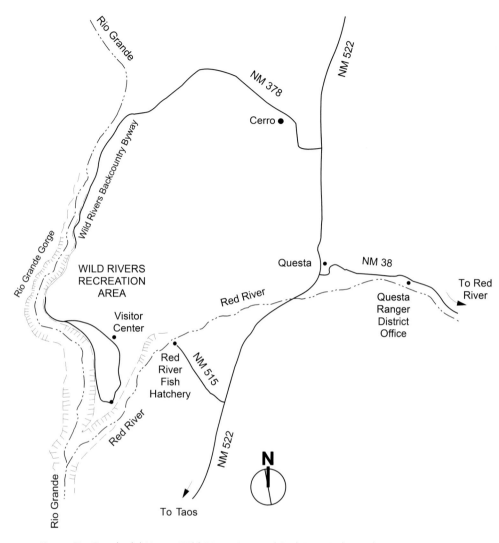

Map 7. Rio Grande del Norte: Wild Rivers Area and Red River Fish Hatchery

(a tributary of the Rio Grande) lined with willows and other riparian habitat. Some of the land accessible from the hatchery is included in the Rio Grande del Norte National Monument: Wild Rivers Area, managed by the Bureau of Land Management. Begin your birding as you drive west along NM 515, inspecting the trees along the highway and at picnic areas. While you might be visiting for the birds, it is worthwhile to look at the exhibits explaining the trout-rearing process when you arrive at the hatchery Visitor Center.

The best birding is located on the east side of the river. Spend some time on the bridge that leads to the hatchery and nearby, searching the rocks for American Dipper. Walk along the path that leads from the parking area to a footbridge over the river, again stopping on the bridge to survey for American Dipper. Cross over the bridge and walk along the fishing trail for a while, although it can be uneven and muddy in locations. When you return to the parking area, take a few minutes to scan the children's fishing pond.

The area around the hatchery is open 8:00 a.m. to 5:00 p.m., and the fishing access area is open 24 hours daily.

County: Taos

eBird Hotspots: Red River Fish Hatchery

Target Birds

Belted Kingfisher It is present year-round and can be seen and heard as it perches in a tree or flies over the river or pond.

Red River Fish Hatchery

Steller's Jay and Western Scrub-Jay While the Steller's Jay is more likely to be seen in the ponderosa pines near the river, the Western Scrub-Jay is found in piñon-juniper habitat along NM 515 and around the hatchery.

Violet-green Swallow It is prevalent during the summer months.

American Dipper It can be seen year-round and has been known to nest under one of the bridges. Look for it bobbing on a rock or flying low along the river.

Townsend's Solitaire It arrives from higher elevations in early October and is present through the end of April.

Song Sparrow It is possible to see Song Sparrow year-round in the willows near the river.

Other Birds

During winter months, check for Canada Goose, Green-winged Teal, and Common Merganser on the river. Other species to look for during the winter include Hairy Woodpecker, Mountain Chickadee, European Starling, and Dark-eyed Junco.

During migration, keep your eyes open for Osprey, Wilson's Snipe, and Tree Swallow.

Species that can be seen during the summer include Turkey Vulture, Common Nighthawk, Broad-tailed Hummingbird, Red-naped Sapsucker, Northern Flicker, Western Wood-Pewee, Cordilleran and Ash-throated Flycatchers, Plumbeous Vireo, Rock Wren, Virginia's and Yellow Warblers, and Chipping Sparrow.

Birds that can be seen at any time of year include Mallard, Red-tailed Hawk, Black-billed Magpie, American Crow, Common Raven, American Robin, Juniper Titmouse, Yellow-rumped Warbler, and Red-winged Blackbird. Other year-round possibilities that are more transient include Pinyon Jay, Clark's Nutcracker, Pygmy Nuthatch, Bewick's Wren, and Cedar Waxwing.

DIRECTIONS

From Taos: Travel 4.1 miles north on US 64W to NM 522. Take NM 522 north 16.5 miles to signs for Red River Fish Hatchery on the left (NM 515).

From Questa: At the intersection of NM 522 and NM 38, travel 3.6 miles south on NM 522 to the sign for Red River Fish Hatchery and NM 515 on the right.

PARKING

There is a large parking area next to the hatchery. There is also adequate parking adjacent to the fishing area.

FEES

None

SPECIAL CONSIDERATIONS AND HAZARDS

See chapter 2 for further safety guidelines.

- Poison ivy: It is found at locations along the Red River.
- Black bears: Although not regular, bears may be more prevalent during drought conditions.
- Cell phone service: There is no cell phone coverage in this area.
- Trail to confluence: It is unsafe to follow the trail on the west side of the Red River to its confluence with the Rio Grande, as the trail disappears after about 2 miles.

FACILITIES

- Accessibility: Areas around the hatchery Visitor Center and the east side of the river between the two bridges are level. The fishing trail on the west side of the river is rocky and uneven. The initial portion of the 1.7-mile Pescado Trail that leads up to the Visitor Center area of the Wild Rivers section of the Rio Grande del Norte National Monument is steep and rocky.
- Restrooms: Visitors can use restrooms inside the hatchery Visitor Center when it is open.
- Water: There is a drinking fountain in the hatchery Visitor Center. If visiting when the Visitor Center is not open, bring sufficient water.
- Picnic tables: There are picnic areas along NM 515.
- Pets: Keep dogs on leash.

CAMPING

Primitive camping is available along NM 515. More established campsites can be found in the Rio Grande del Norte National Monument: Wild Rivers Area.

GAS, FOOD, AND LODGING
Gas, food, and lodging are available in Questa.

Rio Grande del Norte National Monument: Wild Rivers Area
Description

The Wild Rivers Area of the Rio Grande del Norte National Monument is located at the confluence of the Rio Grande and the Red River. The junction of the two rivers can be viewed from La Junta Point high above the vast canyon. The area, administered by the Bureau of Land Management, is at an elevation of 7,500 feet and encompasses a variety of habitats. It is part of the Upper Rio Grande Gorge Important Bird Area (IBA), one of Audubon New Mexico's Priority IBAs. The 800-foot-deep river gorge—part of the Rio Grande rift—cuts through volcanic rock and provides nesting habitat for raptors. A 13-mile road, the Wild Rivers Backcountry Byway, makes a loop on the Taos Plateau (Great Basin scrub, grassland, and piñon-juniper habitat), providing access to scenic overlooks, trails, and camping areas. Some of the trails leading to the river descend into ponderosa pine habitat. In addition to looking for birdlife, keep your eyes peeled for Rocky Mountain bighorn sheep on the steep cliffs.

There are a variety of ways to experience the natural resources of the Wild Rivers Area, some leisurely and some strenuous. Start your birding as you drive west on NM 378. The agricultural area around the community of Cerro can often yield Black-billed Magpie. As you drive along the Backcountry Byway, take the time to stop at each of the overlooks to scan the thermals and rocky cliffs for raptors and to listen for Rock and Canyon Wrens.

The Chiflo Trail (0.4 mile one way), on your right shortly after entering the Wild Rivers Area, descends into the canyon and has a gradual elevation change. Hiking poles may be useful as its surface is somewhat rocky. Near the river during the summer, you might encounter Yellow Warbler or Yellow-breasted Chat. The Little Arsenic (1.0 mile one way—moderate elevation change) and La Junta (1.2 miles one way—difficult, rocky switchbacks along the vertical cliffs, a ladder, and steep stairs) Trails also lead into the gorge and provide habitat for migrating warblers and nesting passerines.

The Pescado Trail can be accessed from the Visitor Center or from the Red River Fish Hatchery. The only difficult part is a very short section just above the hatchery that is rather steep and rocky. The beauty of this trail is

Confluence of Red River and Rio Grande

that it traverses several habitats, including open fields, piñon-juniper, ponderosa pine, and riparian at its terminus at the Red River. Round-trip is 3.4 miles, but even shorter walks starting at either end can be rewarding.

The East Rim and Rinconada Loop Trails and the Rio Bravo Nature Trail, in piñon-juniper habitat, provide opportunities for level walking and birding along the canyon rim.

The Wild Rivers Area is open year-round for camping and day use. The Visitor Center is open 10:00 a.m. to 3:00 p.m. from Memorial Day to Labor Day and on major holiday weekends.

County: Taos
eBird Hotspots: Wild Rivers Recreation Area
website: http://www.blm.gov/nm/st/en/prog/recreation/taos/wild
_rivers_rec_area.html

Golden Eagle This majestic raptor maintains a year-round territory in the area of Wild Rivers and often can be seen hunting on the thermals.

Red-tailed Hawk It is present year-round and may be seen flying or perched on the top of a juniper.

Pinyon Jay While it is not consistently seen, flocks of this roving corvid often swoop in and land on the top of a piñon pine to feed briefly before flying off.

Juniper Titmouse It is often heard before seen. Locate it by listening for its soft tapping or trill among piñon or juniper.

Rock and Canyon Wrens Both wrens live in the canyons. The Rock Wren can be found near crevices in the rocks, where it forages for insects, or bobbing on top of a boulder. The Canyon Wren also makes its home in rock clefts and can be camouflaged by the similarly colored sandstone. Its descending song often can be heard echoing in the canyon.

Mountain and Western Bluebirds Both bluebirds are present year-round, although the Western Bluebird is more prevalent.

Other Birds

While the winter weather can be challenging, visiting during the winter might yield a Bald Eagle over the rivers or a Great Horned Owl on a ledge of the gorge. In addition to the target birds, look for Downy and Hairy Woodpeckers, Black-capped and Mountain Chickadees, Townsend's Solitaire, and Dark-eyed Junco.

Pinyon Jay (photo by Mouser Williams)

Peregrine Falcon is known to nest in the canyon, arriving in early spring.

During the summer you can find Turkey Vulture; Mourning Dove; White-throated Swift; Black-chinned and Broad-tailed Hummingbirds; Northern Flicker; American Kestrel; Gray and Ash-throated Flycatchers; Western Wood-Pewee; Say's Phoebe; Plumbeous Vireo; Violet-green Swallow; Blue-gray Gnatcatcher; American Robin; Sage Thrasher; Yellow-rumped and Black-throated Gray Warblers; Rock Wren; Spotted Towhee; Rufous-crowned, Brewer's, Chipping, Vesper, Lark, Sagebrush, Savannah, and Song Sparrows; Brown-headed Cowbird; and Black-headed Grosbeak.

Migration can bring Canada Goose, Mallard, Blue-winged Teal, and Northern Shoveler on the rivers. Rufous and Calliope Hummingbirds migrate through the area midsummer through midfall. Other species seen during migration include Red-naped Sapsucker (spring), Osprey, Olive-sided Flycatcher, Bewick's Wren, Ruby-crowned Kinglet, Green-tailed Towhee, and Western Tanager.

Year-round species include Steller's Jay, Western Scrub-Jay, Common Raven, Canyon Wren, Juniper Titmouse, Bushtit, and Canyon Towhee.

DIRECTIONS

From Taos, travel 4.1 miles north on US 64W to NM 522. Take NM 522 north approximately 22.5 miles and turn left on NM 378 toward the community of Cerro (NM 378 begins 2.5 miles north of the town of Questa). After a quarter mile the road bears right, then left through Cerro. Continue on NM 378 as the road loops south along the canyon rim to the Visitor Center (approximately 12 miles from NM 522).

PARKING

There is parking at all day-use areas. Campgrounds also have day-use parking areas.

FEES

Day-use fees are charged at all developed sites at Wild Rivers; day-use passes are available from BLM, and federal passes waive day-use fees.

SPECIAL CONSIDERATIONS AND HAZARDS

See chapter 2 for further safety guidelines.

- Steep, vertical cliffs: Avoid walking close to the canyon rim.
- Rattlesnakes: Rattlesnakes are possible. Stay on established trails and avoid reaching into rocky ledge crevices if hiking down to the river.
- Black bears: While not prevalent, black bears are possible.
- Cougars: Cougars may be present, particularly in the early morning and in the evening.
- Cultural resources: The area served as a corridor for travel for many centuries; archaeological resources, specifically petroglyphs, have endured and are protected under the Archaeological Resources Protection Act.
- Freshwater springs: No swimming is allowed to protect water quality.
- Weather considerations: The area is subject to summer thunderstorms and can experience severe winter weather. It is advisable to check the New Mexico Department of Transportation website (www.nmroads.com) or call the hotline (800–432–4269) for current road conditions during winter months.

FACILITIES

- Accessibility: There are two wheelchair-accessible overlooks (La Junta Point and Chawalauna Overlook); other overlook areas are graded and provide level walking. Three trails (Chiflo, Rinconada Loop, and East Rim) are level, easy walking (a few with rocky surfaces may necessitate the use of a walking stick for some people). Other trails are rated from moderate to difficult, with elevation gains/drops from 500 to 1,000 feet on steep trails, requiring climbing stairs or over jagged volcanic rocks.
- Restrooms: Accessible restrooms are located at picnic areas and campgrounds.
- Water: Available at selected campgrounds
- Picnic tables: There are picnic tables at a number of locations.
- Pets: The National Monument rules require that dogs be on leash and controlled. Pets are not allowed on some of the trails.

CAMPING

There are several campgrounds along the rim and a few along the River Trail adjacent to the Rio Grande.

Gas, food, and lodging are available in Questa.

Along NM 38: Questa to Eagle Nest
Description

NM 38 extends from the Village of Questa and follows the Red River Canyon to the town of Red River and then continues east over Bobcat Pass, where it descends into the Moreno Valley at Eagle Nest. This route provides treasures at each season and is often highlighted as having winter snow vistas, fields of summer wildflowers, and spectacular autumn leaf viewing. A number of locations along this route are recommended to bird in a variety of habitats.

Questa Ranger District Offices (Carson National Forest): Just prior to arriving at the ranger station, you will pass Eagle Rock Lake. The lake, river, and nearby tailings ponds were added in 2011 to the Superfund National Priorities List for federal cleanup of contaminants from the molybdenum mining operation in the canyon. Eagle Rock Lake currently does not attract waterfowl. There is good birding in the area surrounding the district offices (elevation 7,480 feet), which border the Red River on the north. A service road on the south supports riparian species.

Columbine Campground: The highway passes through the massive mining area before arriving at Columbine Campground, 5 miles east of Questa at the mouth of Columbine Canyon at 7,900 feet. Although the Columbine Trail (#71) extends for 14.2 miles, the first 1.25 miles pass through mixed conifer habitat, are fairly level, and provide opportunities for middle-elevation species.

Town of Red River: The Red River Nature Trail at 8,700 feet is a 2-mile interpretive trail that runs east along the river starting at the Platinum Chairlift on Pioneer Road and ending at Goose Lake Road. The chairlift also operates daily in summer and provides opportunities for higher-altitude birding. If you visit during the winter, look for homes with feeders, as all three species of rosy-finch often can be seen.

Enchanted Forest Cross-Country Ski Area: Located 4 miles east of Red River at 9,788 feet, the area offers prospects for year-round birding in spruce-fir habitat. A network of cross-country and snowshoe trails are open to the public for hiking when there is no snow present. During snow season you can bird along the road into the cross-country ski area (part of Carson National Forest) and around the parking lot.

View from Enchanted Forest Cross-Country Ski Area along NM 38

Shortly after the cross-country ski area, NM 38 leaves Taos County at the national forest border and enters Colfax County. As you descend into the Moreno Valley and Eagle Nest, look to your right for the all-but-abandoned ruins of Elizabethtown, an old gold mining town and home to more than 7,000 people in 1870.

Counties: Taos and Colfax

eBird Hotspots: Red River Nature Trail and Enchanted Forest Cross-Country Ski Area

Target Birds

Broad-tailed Hummingbird It can be seen all along this route from mid-April through the end of September.

Clark's Nutcracker (photo by Jim Tuomey)

Williamson's Sapsucker Found at all sites along this route, it is most likely to be seen in July.

Cordilleran Flycatcher This high-elevation flycatcher arrives mid-May and is present through the end of August. A good place to look for one is near structures, such as picnic shelters and cabin porches.

American Dipper It is possible to spot one bobbing on the rocks anywhere along either the Red or Moreno River.

Clark's Nutcracker This corvid can be found at the Enchanted Forest Cross-Country Ski Area or along the highway nearby.

Rosy-finches All three species of rosy-finch visit feeders in Red River and often can be seen foraging near the Enchanted Forest Cross-Country Ski Area.

Cassin's Finch This high-altitude finch is possible year-round. It is often at feeders in the town of Red River.

Evening Grosbeak While it is possible at any time of year, it moves around depending on food supply.

Other Birds

Year-round species include Hairy Woodpecker; Gray (Enchanted Forest Cross-Country Ski Area) and Steller's Jays; Black-billed Magpie; American Crow; Common Raven; Black-capped (ranger station) and Mountain Chickadees; Red-breasted, White-breasted, and Pygmy Nuthatches; Brown Creeper; Mountain Bluebird; and Dark-eyed (Gray-headed) Junco.

During the summer, look for Mallard (along the Red River), Red-tailed and Cooper's Hawks, Turkey Vulture, Black-chinned Hummingbird (ranger station), Red-naped Sapsucker, Downy Woodpecker (ranger station), Northern Flicker (can be seen in winter at ranger station), Western Wood-Pewee (ranger station), Warbling Vireo, Violet-green Swallow, Ruby-crowned Kinglet (Enchanted Forest Cross-Country Ski Area and Red River), American Robin, Yellow (ranger station) and Yellow-rumped Warblers, Green-tailed Towhee (Red River), Chipping and Lincoln's Sparrows (Enchanted Forest Cross-Country Ski Area and Red River), Western Tanager (only at ranger station during migration), and Black-headed Grosbeak.

During migration it is possible to see Band-tailed Pigeon, White-crowned Sparrow, American Goldfinch, and Rufous and Calliope Hummingbirds (late summer to fall).

In winter, watch for Wild Turkey on south-facing slopes along the road. In addition to the rosy-finches, additional subspecies of Dark-eyed Junco come to feeders at Red River, including Oregon, Slate-colored, and Pink-sided.

DIRECTIONS

The western end of NM 38 begins at Questa (approximately 25 miles north of Taos via US 64 and NM 522). From NM 522 at Questa, turn east onto NM 38. The Questa Ranger District Office (ranger station) is approximately 1.5 miles east of the village. Columbine Campground is about 5 miles east of Questa. Continue on to the town of Red River (about 12 miles beyond Questa). The turnoff (Sangre de Cristo Road) to Enchanted Forest Cross-Country Ski Area is approximately 16 miles east of Questa. The eastern end of NM 38 is at the village of Eagle Nest in the Moreno Valley, 29 miles from Questa.

PARKING

Parking is available at the ranger station, Columbine Campground at the trailhead, ski area lots in Red River, the turnout along NM 38 at Sangre de Cristo Road, or in the parking lot of Enchanted Forest Cross-Country Ski Area.

FEES
There is a day-use fee for the picnic area at Columbine Campground but no fee for parking at the trailhead. There is a fee for using the summer chairlift at Red River. No fee is charged for hiking at Enchanted Forest Cross-Country Ski Area.

SPECIAL CONSIDERATIONS AND HAZARDS
See chapter 2 for further safety guidelines.

- Winter travel: NM 38 can be snowpacked and icy during the winter. It is advisable to check the New Mexico Department of Transportation website (www.nmroads.com) or call the hotline (800–432–4269) for current road conditions during winter months.
- All-terrain vehicles, mountain bikes, and runners: Many backcountry trails are designated multiuse. Watch and listen for vehicles on these trails.
- Black bears: Bears are possible at Columbine Campground and Trail and at the Enchanted Forest Cross-Country Ski Area, especially if drought conditions exist. Do not leave food unattended.
- Cougar: Cougars are possible in the high country, especially if you are hiking early in the morning or at dusk.
- Poison ivy and stinging nettle: Both plants are possible along the Red River.
- Cell phone service: Cell phone coverage is spotty or not available along this route.

FACILITIES
- Accessibility: With the exception of trails in the Enchanted Forest Cross-Country Ski Area, most trails are fairly level. At the ski area, some trails are uneven and might require a walking stick.
- Restrooms: Columbine Campground (May–October); in Red River only at public buildings (when open); none at Enchanted Forest Cross-Country Ski Area
- Water: Drinking fountain in ranger station and Columbine Camp-

ground (May–October); may be purchased in Red River; none at En-
chanted Forest Cross-Country Ski Area
- Picnic tables: Columbine Campground (May–October) and Mal-
lette Park in Red River; none at Enchanted Forest Cross-Country Ski
Area
- Pets: Carson National Forest rules require that dogs be on leash in
Columbine Campground. There are separate trails for leashed dogs at
Enchanted Forest Cross-Country Ski Area.

CAMPING
Camping is available at Columbine Campground and numerous RV parks
in Red River. At Enchanted Forest Cross-Country Ski Area backcountry yurt
rentals are available year-round.

GAS, FOOD, AND LODGING
Gas, food, and lodging are available in Questa, Red River, and Eagle Nest.

Cimarron Canyon State Park
Description
Sheltered in an 8-mile-long wooded canyon, bordered by rugged hills on the
west and palisade-like cliffs along the eastern end, the state park and surrounding Colin Neblett Wildlife Management Area lives up to its name,
which in Spanish means "wild and untamed." The Cimarron River flowing
out of Eagle Nest Lake below the dam tumbles through the canyon and is a
good location for American Dipper. Wild Turkey is abundant and can be
seen almost anywhere within the park. During the winter, you might see a
Bald Eagle perched in a ponderosa pine along the river.

Traveling east from the village of Eagle Nest on US 64, the road passes
over the Cimarron Range and then drops down into the canyon, where you
enter the state park. The first stop on the right is the loop trailhead for Tolby
Creek Canyon.

The Tolby Creek Trail, an old logging road (11 miles round-trip, steep,
rocky, and wide), is located in the Colin Neblett Wildlife Management Area
and climbs up to 9,000 feet to an aspen-ringed meadow where Dusky
Grouse is possible, as well as high-mountain meadow species, such as
Green-tailed Towhee. If you do not want to hike the trail, take time to bird
the perimeter of the trailhead parking area, particularly along Tolby Creek,

Tolby Creek Trail, Cimarron Canyon State Park

which runs along the western boundary of the trailhead. There is free parking at the trailhead with a park pass; however, a GAIN (Gaining Access into Nature) permit is needed to hike the trail (see chapter 2 for more information).

Tolby Day-Use Area is the next stop, across the highway adjacent to the campground and state park Visitor Center. Explore the willows and shrubs surrounding the river by walking along the various informal fishing trails. This is particularly productive in the summer and during migration.

Drive east down the canyon to the Special Trout Waters (catch and release). Look for mile marker 290. Parking is available at a turnout on your right 0.6 mile beyond the mile marker sign. Walk through the open area along the fishing trails for the possibility of spotting sapsuckers (Red-naped and Williamson's), flycatchers, vireos, Gray Catbird in summer, and Red-breasted Nuthatch in fall.

At mile marker 292 there are turnouts on both sides of the road. The trailhead for the Clear Creek Trail (3.5 miles round-trip into the Colin Neblett Wildlife Management Area) is on the right. The trail follows Clear Creek and crosses several log bridges. A short section of the trail is steep and narrow. The trail ends at the third waterfall and returns along the same route. Sapsuckers nest in the aspens, as well as Ruby-crowned Kinglet and Brown Creeper. The area on the north side of the highway is another good location to check for American Dipper. There is free parking at the trailhead with a park pass; however, a GAIN permit is needed to hike the trail.

The next stop is the Blackjack Day-Use Area, where you can explore the fishing trails along the river.

When you see the towering volcanic cliffs in long columns on your left, look for the Palisades Picnic Area, which will be on your right. Watch for White-throated Swift and swallows below the cliffs and bird the picnic area.

There is a day-use parking area in the Maverick Campground. Walk over to the Gravel Pit Lakes and explore the shoreline and willows surrounding the lake.

The state park is open year-round; however, there are some seasonal closures in parts of the park. Colin Neblett, Tolby Creek, and Maverick Trails are closed May 15–July 31; Clear Creek Trail is open year-round. Day-use areas are open 6:00 a.m. to 9:00 p.m.

County: Colfax

eBird Hotspots: Cimarron Canyon State Park and Cimarron Canyon State Park—Tolby Trail/Campground

website: http://www.emnrd.state.nm.us/SPD/cimarroncanyonstatepark.html

Target Birds

Wild Turkey At any time of year, you are likely to encounter Wild Turkey as you explore the park.

Northern Goshawk While not guaranteed, this is a good location to spot this montane accipiter year-round, particularly on the Tolby Creek Trail.

Red-naped Sapsucker It breeds in the canyon. Look for it in ponderosa pine and aspens.

Cordilleran Flycatcher It can be found in the park during the summer months, arriving by mid-May and migrating by late August.

American Dipper The dipper loves the cascading Cimarron River and might be seen anywhere along the river.

Gray Catbird This is a good location to find this species in the underbrush along the river and streams.

Virginia's Warbler It prefers dense scrub understory and often can be seen on the ground. It arrives late April/early May and is possible through August.

MacGillivray's Warbler It nests in the low shrubs along the river. If you walk the fishing trails in the Tolby Day-Use Area during the summer, you should find one or more.

Green-tailed Towhee Look for it in scrubby habitat at forest edges in the park.

Western Tanager This stunning tanager arrives in the canyon in mid-May and is present through mid-September. Look for it on hillsides in mixed conifer habitat.

Other Birds

During the winter it is possible to see Golden and Bald Eagles, Townsend's Solitaire, and White-crowned Sparrow.

During migration, look for Band-tailed Pigeon; Hammond's Flycatcher; Rufous Hummingbird; and Yellow, Orange-crowned, Townsend's (fall), and Wilson's Warblers.

Summering species include Turkey Vulture; White-throated Swift; Black-chinned and Broad-tailed Hummingbirds; Williamson's Sapsucker; Olive-sided Flycatcher; Western Wood-Pewee; Warbling Vireo; Tree, Violet-green, Barn, and Cliff Swallows; Ruby-crowned Kinglet; House Wren; Virginia's and Yellow-rumped Warblers; Chipping and Song Sparrows; and Black-headed Grosbeak

Birds that can be seen at any time of year include Red-tailed Hawk, Great Horned Owl, Belted Kingfisher, Northern Flicker, Steller's Jay, Black-billed Magpie, Clark's Nutcracker, American Crow, Common Raven, Black-capped and Mountain Chickadees, Red-breasted and White-breasted Nuthatches, Brown Creeper, Hermit Thrush, American Robin, Dark-eyed Junco, and Evening Grosbeak (irregular).

DIRECTIONS

From Taos: Drive east on US 64 (Kit Carson Road, one block east of the Taos

Plaza) approximately 31 miles, ascending the Sangre de Cristo Mountains to the village of Eagle Nest. At Eagle Nest, US 64 continues east over the Cimarron Range for 3 miles to the Cimarron Canyon State Park boundary.

From Questa and Red River: Drive east from Questa on NM 38 approximately 29 miles through the town of Red River, continuing over Bobcat Pass and south into the Moreno Valley to the village of Eagle Nest. At Eagle Nest turn right (east) onto US 64 and travel east 3 miles to the Cimarron Canyon State Park boundary.

From Las Vegas, New Mexico: At the intersection of I-25 and NM 104, drive 73 miles north on I-25 to NM 58 (Exit 419). Take NM 58 west, traveling approximately 20 miles to the junction with US 64 (at the town of Cimarron). Continue west on US 64 approximately 13.5 miles to the Cimarron Canyon State Park boundary.

PARKING

Parking is available in day-use areas, as well as at a number of turnouts along the river.

FEES

The fees at this site are complicated because of the interwoven properties of the New Mexico State Parks and the New Mexico Department of Game and Fish.

- New Mexico State Parks day-use fee or annual pass is needed to park in or use any of the day-use areas or trailhead parking areas.
- GAIN and HMAV are required to hike on the Tolby Creek, Clear Creek, and Maverick Trails. See chapter 2 for more information about these passes.

SPECIAL CONSIDERATIONS AND HAZARDS

See chapter 2 for further safety guidelines.
- Hunting season: Tolby Creek and Maverick Trails are used by hunters in season.
- Mosquitoes: Mosquitoes are possible along the river during the summer.
- Black bears: While bears are possible at all times, during drought years the bears' food supplies may be more limited, causing them to wander into more populated areas. Do not leave food unattended.

- Cougars: Cougars may be present, particularly in the early morning and in the evening.
- Stinging nettle: This plant grows along the river in shaded areas and can be unnoticed until brushed against.
- Poison ivy: Poison ivy can be found along the river and, in some areas, along the roadside.
- Small mammals and plague: Avoid contact with small mammals, such as squirrels and chipmunks. If you are traveling with a dog, be sure to keep the dog on leash to avoid contact with an infected rodent.
- Cell phone service: There is no cell phone coverage in most areas of the park.
- Sensitive geologic area: Rock climbing is prohibited in the Palisades area to protect the towering igneous rock cliffs.
- Winter driving conditions: At times, there are winter road advisories. It is advisable to check the New Mexico Department of Transportation website (www.nmroads.com) or call the hotline (800–432–4269) before driving into the area during the winter.

FACILITIES

- Accessibility: There are accessible parking and restrooms in the campgrounds and day-use areas. Some of the fishing trails that lead along the river can be uneven. The Tolby Creek and Clear Creek Trails are steep and uneven in areas.
- Restrooms: Tolby Day-Use Area, Blackjack Day-Use Area, Palisades Picnic and Day-Use Area, as well as in campgrounds
- Water: Available at Tolby, Maverick, and Ponderosa Campgrounds
- Picnic tables: Tolby, Maverick, and Ponderosa Campgrounds
- Pets: Pets must be on leash in the state park and Colin Neblett Wildlife Management Area.

CAMPING

Several campgrounds are scattered throughout the state park.

GAS, FOOD, AND LODGING

Food is available in Eagle Nest. Gas and lodging are available in Eagle Nest and Ute Park, just east of the park.

Eagle Nest Lake

Eagle Nest Lake State Park
Description
Located at an elevation of 8,300 feet in the historic Moreno Valley, the village and state park are flanked by two of New Mexico's highest peaks: Baldy Mountain (12,441 feet) to the north and Wheeler Peak (13,161 feet) to the west. Eagle Nest Lake State Park borders the Colin Neblett Wildlife Management Area (WMA) on the east. While the 2,200-acre lake (reservoir) at the headwaters of the Cimarron River is well known by anglers, it also attracts a variety of waterbirds, including summering Common Merganser and White Pelican. In addition, there are a variety of raptors, including Golden Eagle, as well as other high-elevation land birds.

Stop first at the Visitor Center area to pick up a park map with directions to each of the day-use areas and to view the displays. The campground is closed from December through mid-March, but the rest of the area is open year-round. The shore near the boat ramps provides good viewing of the lake, including the inlet where waterbirds gather near the mouth of Moreno Creek.

Two free day-use areas are worth exploring. Both are open year-round, weather permitting. The Moreno Day-Use Area at the north end of the lake also provides good viewing areas of the lake. It is a short walk from the Six-Mile Day-Use Area to the lake. The Cieneguilla Day-Use Area (fee) on the south end of the lake is open April–October. It provides opportunities to bird along Cieneguilla Creek. As you drive into both Six-Mile and Cieneguilla Day-Use Areas, look for Mountain Bluebird and Brewer's Blackbird in the fields on either side of the road.

The two branches of the Lake View Trail lead through the grasslands. One branch starts in town south of Moreno Creek and ends at the Visitor Center (0.4 mile), and the other branch starts north of Moreno Creek and ends at the Moreno Day-Use Area (1.2 miles). The grassland can yield Savannah and Vesper Sparrows in season.

County: Colfax

eBird Hotspots: Eagle Nest SP

Web site: http://www.emnrd.state.nm.us/SPD/eaglenestlakestatepark .html

Target Birds

Common Merganser It is present from mid-March through the end of October.

Western and Clark's Grebes Both of these grebes are present from early May through October. There has been only one record of breeding.

American White Pelican Eagle Nest Lake is the most reliable location to view this species in northeastern New Mexico. While it can be seen from late May through the end of September, the largest concentrations occur between the end of June and mid-July when they use Eagle Nest Lake as a staging area during migration.

California Gull Be sure to scan the flocks of Ring-billed Gulls for California Gull. Several can usually be spotted between the end of March and the end of October.

Mountain Bluebird This striking bluebird can be found in the grassy areas surrounding the lake, as well as in the village of Eagle Nest from mid-March through the end of October.

Savannah Sparrow It arrives mid-May and can be seen in the grasslands around the lake until the end of September.

Other Birds

During the winter months (November–March) when the lake is not frozen, look for Green-winged Teal. Other winter birds include Wild Turkey, Rough-legged Hawk, and American Tree Sparrow. Keep your eyes open in town—during some winters, all three species of rosy-finch will descend into the valley and visit feeders.

During the summer you can find Gadwall; American Wigeon; Blue-winged, Cinnamon, and Green-winged Teal; Pied-billed, Western, and Clark's Grebes; Double-crested Cormorant; Great Blue Heron; Bald Eagle; American Coot; Killdeer; and Ring-billed Gull on or near the lake. Other summer species include Turkey Vulture; Northern Harrier; Broad-tailed Hummingbird; Northern Flicker; Northern Rough-winged, Tree, Violet-green, Barn, and Cliff Swallows; American Robin; Vesper Sparrow; Red-winged Blackbird; Western Meadowlark; Brewer's Blackbird; and Common Grackle.

Migrating waterfowl that use the lake include Northern Shoveler, Northern Pintail, Canvasback, Redhead, Bufflehead, and Ring-necked and Ruddy Ducks. Other migrating species include Ferruginous Hawk, Franklin's Gull, Rufous Hummingbird, Western Bluebird, American Pipit, Yellow-rumped Warbler, Chipping Sparrow, and Yellow-headed Blackbird.

Year-round birds at the state park and surrounding area include Canada Goose, Mallard, Red-tailed Hawk, Hairy Woodpecker, Steller's Jay, Black-billed Magpie, American Crow, Common Raven, Mountain Chickadee, White-breasted Nuthatch, Horned Lark, European Starling, Dark-eyed Junco, Great-tailed Grackle, and House Sparrow.

DIRECTIONS

From Taos: Drive east on US 64 (Kit Carson Road, one block east of the Taos Plaza) approximately 29 miles over the Sangre de Cristo Mountains and into the Moreno Valley to Eagle Nest Lake State Park Visitor Center entrance road (County Road B11 B, a gravel road) on the right (south of the town of Eagle Nest).

From Questa and Red River: Drive east from Questa on NM 38 approximately 29 miles through the town of Red River, continuing over Bobcat Pass and south into the Moreno Valley to the town of Eagle Nest. At Eagle Nest turn left (west) onto US 64 and travel 1.2 miles to Eagle Nest Lake State Park Visitor Center entrance road (CR B11B) on the left (south of the village of Eagle Nest).

From Las Vegas, New Mexico: At the intersection of I-25 and NM 104, drive 73 miles north on I-25 to NM 58 (Exit 419). Take NM 58 west, traveling approximately 20 miles to the junction with US 64 (at the town of Cimarron). Continue west on US 64 through Cimarron Canyon State Park and the village of Eagle Nest approximately 25.5 miles to the Eagle Nest Lake State Park Visitor Center entrance road (CR B11B) on the left.

PARKING

Parking is available at several areas around the lake. Parking for the Lakeview Trail is along the street in town.

FEES

There is no fee at the two day-use areas (Moreno, Six-Mile) or to walk on Lakeview Trail. Day-use fee or state park annual pass is required at Cieneguilla Day-Use Area.

SPECIAL CONSIDERATIONS AND HAZARDS

See chapter 2 for further safety guidelines.

- Black bears: Bears may be attracted to picnic or camping areas and are sometimes found in town.
- Cougars: They have been found on the east side of the lake (Colin Neblett WMA).
- Prairie dog tunnels: Look carefully when walking on trails. Prairie dog tunnels are very close to the surface and may collapse with foot pressure.
- Plague: Area rodents have been infected, which is a potential hazard to pets.
- Quicksand: Mud along the edge of the lake can act like quicksand—visitors have lost shoes in the mud.
- Blue-green algae blooms: Occurring in summer, the algae can cause skin rash, asthmalike symptoms, vomiting, and nervous system effects. If exposed, rinse skin with fresh water as soon as possible. The algae can be toxic to pets if ingested.
- Winter conditions: The surface of the lake often freezes during the winter, and area roads can be snowpacked and icy. It is advisable to check the New Mexico Department of Transportation website (www.nmroads.com) or call the hotline (800–432–4269) for current road conditions during winter months.

Vietnam Veterans Memorial Chapel

FACILITIES
- Accessibility: Most areas, including the Lakeview Trail, are level, easy walking.
- Restrooms: Moreno Day-Use Area, Visitor Center, and Six-Mile Day-Use Area
- Water: Available in the campground
- Picnic tables: Adjacent to the Visitor Center
- Pets: State park rules require dogs to be on leash.

CAMPING
A campground is located adjacent to the Visitor Center.

GAS, FOOD, AND LODGING
These services can be found in the village of Eagle Nest.

Vietnam Veterans Memorial State Park and Area
Description
Perched on a hillside overlooking the Moreno Valley (at 8,500 feet), Vietnam

Veterans Memorial State Park was established by Victor and Jeanne Westphall to honor their son, Marine First Lieutenant David Westphall, who was killed in Vietnam in May 1968. When it opened in 1971, it was one of the first memorials of its kind in the United States dedicated to Vietnam veterans. The grounds are planted and maintained to foster a healing environment for visiting Vietnam veterans and their families. This landscaping and preservation of the natural habitat attract a wide variety of birds year-round.

As you walk from the parking area to the Visitor Center, stop at the fountain and spend some time on one of the benches watching the myriad birds that visit from spring through fall.

Follow the path to the Visitor Center and gift shop area. This is a good location to scan the adjacent valley for raptors hunting for prairie dogs—one of the highlights of birding at this location. Elk are often present in the valley as well.

The park and Reflection Room are open 24 hours daily. The Visitor Center is open 9:00 a.m. to 5:00 p.m. daily (April 1–October 31). The Visitor Center is closed Tuesdays and Wednesdays (November 1–March 31).

After birding at the state park, return to US 64 and turn left. In approximately 1.3 miles you will come to a dirt road—County Road B36 (Saladon Road). While the first part is a county road, a sign will indicate when the road becomes private. Although there are no formal turnouts, you can pull to the side of the road to allow other vehicles to pass. Do not block the entrance to a private drive, and remain near your vehicle in case you need to move it. The road passes through pastures where Great Blue Heron, Horned Lark, Barn Swallow, Mountain Bluebird, Dark-eyed Junco, and Brewer's Blackbird are possible. Scan the nearby small quarry, the hills to the east, and the cottonwood trees in the pasture for both Golden and Bald Eagles. Other raptors you might encounter include Northern Harrier; Swainson's, Red-tailed, and Ferruginous Hawks; and Prairie Falcon.

County: Colfax

eBird Hotspots: Vietnam Veterans Memorial SP

website: http://www.emnrd.state.nm.us/SPD/vietnamveteransmemorial statepark.html

Target Birds

Golden Eagle It nests on the cliffs in the Cimarron Mountains on the east side of the Moreno Valley across from the state park and hunts in the valley below. It often can be seen perched in the cottonwood trees viewed from CR B36. In spring, immature Golden Eagles may be seen practicing flying skills in the updrafts at cliffs near the state park or CR B36.

Bald Eagle Although it does not breed in the valley, it is present year-round (scarce in winter). It also can be seen hunting in the valley across from the state park.

Red-tailed Hawk When not soaring overhead or hunting in the arroyo south of the state park or in the valley to the east, it often perches on fence posts or power poles along US 64.

Ferruginous Hawk While usually not very prevalent in the Moreno Valley, it can be seen most often during migration. Look for it along US 64, in trees in the evening on ridgetops on CR B36, or hunting near the state park. Its numbers peak during years when prairie dog populations flourish.

Broad-tailed Hummingbird It is a frequent visitor to the fountain area at the state park.

Horned Lark Look for it in the fields along CR B36 beyond the point where the road turns south.

Mountain Bluebird During breeding, it leaves the nesting area in the conifers on the nearby hillsides and comes to the park fountain to drink and bathe. After breeding, it is more likely found in the pastures along US 64 and CR B36. It is present year-round in the Moreno Valley.

Western Tanager It nests in conifers in the hills west of the state park and comes to the fountain between May and early September.

Other Birds

Summer is the prime viewing time at the state park. Species that visit the area around the fountain include Violet-green Swallow, Western Bluebird, American Robin, Green-tailed Towhee, Black-headed Grosbeak, Brewer's Blackbird, Red Crossbill, Pine Siskin, Lesser Goldfinch, and House Sparrow.

During migration Yellow and Yellow-rumped Warblers visit the fountain. Both Rufous and Calliope Hummingbirds are possible during fall migration.

Raptors that can been seen during migration from both the state park and along the county road include Northern Harrier and Swainson's Hawk.

Species that can be seen at any time of year include Williamson's Sapsucker (in conifers), Hairy Woodpecker, Steller's Jay, Black-billed Magpie, American Crow, Common Raven, Mountain Chickadee, White-breasted Nuthatch, and Evening Grosbeak.

DIRECTIONS

From Taos, one block east of the Taos Plaza at the intersection of US 64 and NM 68, drive east on US 64 approximately 22 miles. This road ascends Palo Flechado Pass in the Sangre de Cristo Mountains and descends into the Moreno Valley just north of the resort town of Angel Fire (south of US 64 on NM 434). At 22 miles from Taos, turn left onto County Road B4 and then immediately right onto Country Club Road for approximately 0.4 mile to the parking lot. The memorial's large, white, sail-like structure will be visible for several miles before the turnoff.

PARKING

There is a large parking lot at the state park.

FEES

None

SPECIAL CONSIDERATIONS AND HAZARDS

See chapter 2 for further safety guidelines.

- Winter road conditions: US 64, CR B36, and the road up to the state park can be snowpacked and icy during the winter. It is advisable to check the New Mexico Department of Transportation website (www.nmroads.com) or call the hotline (800–432–4269) for current road conditions during winter months.

FACILITIES

- Accessibility: The facility is fully accessible.
- Restrooms: Visitor Center
- Water: Drinking fountain located in the Visitor Center
- Picnic tables: There are picnic facilities.
- Pets: Dogs must be kept on leash in the state park and should not run off leash along CR B36.

CAMPING

Camping facilities are available in Angel Fire and at Eagle Nest Lake State Park.

GAS, FOOD, AND LODGING

Gas, food, and lodging are available in the town of Angel Fire.

CHAPTER 6

Jemez Mountains and Los Alamos

General Overview

Not evident unless viewed in a satellite photo or an elevation map, the Jemez Mountains comprise one of the three largest super-volcanoes in the United States active within the last 2 million years. This volcanic activity has resulted in a number of interesting geologic phenomena that form the character of some of the sites in this chapter, including Bandelier National Monument, with its cliffs of solidified pyroclastic flows, and Valles Caldera National Preserve, part of the large central volcanic caldera. In addition to geologic interest, the Jemez Mountains encompass locations of cultural and historical significance.

This chapter includes six sites and routes, starting at the eastern edge of the Jemez Mountains and traveling west. The area includes three major vegetation zones: piñon-juniper, ponderosa, and mixed conifer. Within these zones are montane riparian, alpine meadow, and wetlands/lakes habitats. Bird life is varied and includes one of the few New Mexico sites for nesting Black Swift.

In 2011, the Las Conchas wildfire burned 156,000 acres, and in 2013 the Thompson Ridge fire burned 23,965 acres. These wildfires affected several of the sites in this area. Where applicable, the effect on access, habitat, and resulting bird diversity are addressed in individual site descriptions.

In summer, thundershowers frequently are generated with little warning right above the Jemez Mountains. During the winter, there may be several inches of snow. Daytime temperatures can be close to freezing, and nighttime temperatures may dip into the teens. It is advisable to check the New Mexico Department of Transportation website (www.nmroads.com) or call the hotline (800–432–4269) for current road conditions during winter months.

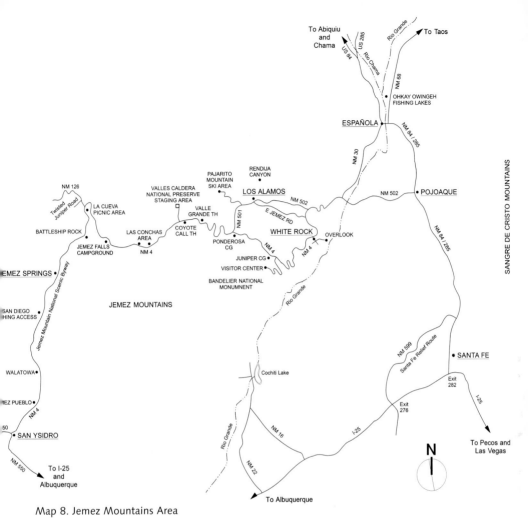

Map 8. Jemez Mountains Area

Bandelier National Monument Visitor Center Area

Description

Nestled in Frijoles Canyon (6,100 feet) on the southeastern edge of the Jemez Mountains below the Pajarito (Spanish for "little bird") Plateau, Bandelier National Monument provides opportunities for birding in several habitats, as well as for visiting the ancient cliff dwellings of the ancestral Puebloan people who inhabited the canyon. The area is designated by Audubon New Mexico as an Important Bird Area.

From the end of May through the end of October, access to the main part of the monument may be required by shuttle bus from White Rock. If you visit at other times of the year, start your birding as you wind your way into Frijoles Canyon from NM 4. Stop at the overlook along the Entrance Road to scan the canyon and check for piñon-juniper species.

Start your visit in the Visitor Center and then head out on the 1.2-mile Main Loop Trail. Both Rock and Canyon Wrens and Canyon Towhee can be found in the rocks in and around the ruins. The trail forks after a short distance. Follow the right-hand trail that passes the Big Kiva and remains of ancient buildings. This part of the trail is accessible both for individuals with mobility limitations and families using strollers. At the far edge of the ground-level ruins, the trail forks again. The left-hand trail loops around and returns to the Visitor Center. The right-hand trail leads up to the cliff dwellings and requires climbing narrow stone stairways, some with limited handrails. At the end of the row of cliff dwellings, the trail follows stone steps down and crosses the canyon floor and Frijoles Creek (bridge may be at risk of washout following flash flooding) to the Nature Trail.

The Nature Trail can also be accessed by crossing the bridge from the parking lot and turning right toward the Cottonwood Picnic Area. The Nature Trail leads north from the end of the paved path through a riparian habitat.

The Falls Trail is accessed by crossing the bridge from the parking lot and turning left. The trailhead is at the end of the paved path. The 1.5-mile (one way) trail is not accessible by persons with mobility issues. It is rocky in places and has a steep descent to the Upper Falls, where it currently ends. As it descends through the canyon, box elders and cottonwood trees provide a dense canopy (and a brilliant golden palate in October). There is a dense understory, including poison ivy and stinging nettle. This understory was decimated during the severe flash floods of 2013, but it is expected to recover.

Be sure to bird the perimeter of the parking lot and picnic area, which can be productive.

The Visitor Center area is open year-round, except January 1 and December 25, from 9:00 a.m. (8:30 a.m. during summer) to 4:30 p.m. These times are subject to change, so please call to check for current information before visiting (505–672–3861). Even when the Visitor Center is closed, visitors can drive to Frijoles Canyon and bird in the early morning and evening.

Counties: Los Alamos and Sandoval

eBird Hotspots: Bandelier NM—Visitor Center and Bandelier NM—Main Loop Trail

Ruins at Bandelier National Monument

website: http://www.nps.gov/band/index.htm
Facebook: https://www.facebook.com/BandelierNPS

Target Birds
Turkey Vulture There is a large Turkey Vulture roost (one of the largest in New Mexico) near the Visitor Center, and up to 80 can be seen in the early morning sunning themselves along the edges of the canyon as they wait for the thermal updrafts. The first ones arrive at the end of March and are present through the end of September/early October. It is a real treat to visit on an early-summer evening and watch them fly in just as the sun sets.
Northern Pygmy-Owl While it is a rare year-round resident in this part of the

monument, most sightings have occurred in late March/early April, just before the breeding season and before leaves appear on the deciduous trees. Because these owls hunt during the day, they are easier to discover than other owls.

White-throated Swift It can be seen flying near and over the tops of the cliffs between late March and the end of September.

Rock and Canyon Wrens Canyon Wren is a year-round resident, and Rock Wren is present only from April through September. Both are found in the rock canyons and ruins. The Canyon Wren, often easier to hear than see, can be located by its descending song.

Townsend's Solitaire It descends from higher elevations and arrives at the end of August/early September and remains through April. Do not be surprised if you hear it singing; it sings outside its breeding territory.

Canyon Towhee It is a year-round resident that might be seen in the scrubby areas along the Main Loop Trail. It also has a penchant for hunting insects under vehicles, so you may see one emerge from under a car in the parking lot.

Hepatic Tanager An uncommon species of the southwestern mountains, it can be found near the Visitor Center or along the Falls Trail where there is ponderosa pine.

Western Tanager This colorful summer resident arrives in May and is present through the end of August.

Other Birds

Year-round birds include Red-tailed Hawk, Great Horned Owl, Williamson's Sapsucker, Downy (uncommon) and Hairy Woodpeckers, Western Scrub-Jay, Common Raven, Mountain Chickadee, Juniper Titmouse, White-breasted and Pygmy Nuthatches, Western Bluebird, American Robin, and House Finch (February–September).

Species that spend the summer include Mourning Dove; Black-chinned and Broad-tailed Hummingbirds; Western Wood-Pewee; Hammond's, Dusky, Gray, and Cordilleran Flycatchers; Say's Phoebe; Ash-throated Flycatcher; Cassin's Kingbird; Plumbeous and Warbling Vireos; Violet-green Swallow; Virginia's and Grace's Warblers; Spotted Towhee; Chipping Sparrow; and Black-headed Grosbeak.

During migration it is possible to see Sandhill Crane flyovers; Rufous Hummingbird (late summer–fall); Olive-sided Flycatcher; House Wren; Yel-

Northern Pygmy Owl (photo by Warren Berg)

low-rumped (Audubon's), Wilson's, and Townsend's (fall) Warblers; and Green-tailed Towhee.

During the winter, look for Red-breasted Nuthatch, Brown Creeper, Ruby-crowned Kinglet, White-crowned Sparrow, and Dark-eyed Junco.

DIRECTIONS

From the intersection of I-25 (Exit 282) and St. Francis Drive (US 84/285) in Santa Fe, follow US 84/285 northwest approximately 19.5 miles to the Los Alamos exit at Pojoaque. Take the Los Alamos Highway (NM 502) west about 12 miles to the junction with NM 4. Turn left (exit ramp will be on the right) and travel approximately 12 miles through the town of White Rock to the monument entrance.

During times when the Visitor Center area is closed to private vehicles (call 505–672–3861 to check for current information before visiting), board the free shuttle bus at White Rock Visitor Center located on the right, 4 miles from NM 502 on NM 4.

PARKING

There is a large parking lot at the White Rock Visitor Center. The parking lot at the monument's Visitor Center is large but can be full at peak times.

FEES

Day-use fee; federal pass waives day-use fee

SPECIAL CONSIDERATIONS AND HAZARDS

See chapter 2 for further safety guidelines.

- Environmental fragility: The area has a fragile ecosystem. It is important to remain on the trails. The area is protected under the Archaeological Resources Protection Act.
- Winter weather: The monument can experience snow during the winter, so the entrance road can be snowpacked and icy. As soon as feasible, snow is removed from the Main Loop Trail. Call 505–672–3861 to check for current information before visiting.
- Summer weather: Thunderstorms often spring up quickly in the afternoon. Keep an eye on the sky as you explore the area. Frijoles Creek can experience flash flooding from storms upstream, even if it is not raining near the Visitor Center.
- Rattlesnakes: Western diamondback rattlesnakes are possible on the Main Loop and likely along the Falls Trail.
- Poison ivy: This low-growing plant can be found along Frijoles Creek.
- Stinging nettle: Keep your eyes open for the bushier, Rocky Mountain variety of stinging nettle along Frijoles Creek.
- Cell phone service: There is no cell phone coverage at this site.

FACILITIES

- Accessibility: The areas around the Visitor Center, parking lot, picnic area, and first part of the Main Loop Trail are fully accessible. See trail descriptions for more details.
- Restrooms: Available next to the Visitor Center and open dawn to dusk

- Water: There is a drinking fountain at the Visitor Center, and bottled water can be purchased in the gift shop.
- Picnic tables: Along Frijoles Creek
- Pets: Pets are not permitted on any of the trails or in the picnic area. With the exception of service animals, pets are not allowed on the shuttle bus. Dogs can be walked along paved roads.

CAMPING

Campsites are available at Bandelier National Monument's Juniper Campground.

GAS, FOOD, AND LODGING

Gas, food, and lodging are available in White Rock and Los Alamos.

Los Alamos: Upper Rendija Canyon and Perimeter Trail

Description

Rendija Canyon is the northernmost valley located between the plateau fingers of volcanic ash deposits that make up the town of Los Alamos and also is part of the Santa Fe National Forest. This site focuses on the upper part of Rendija Canyon, which can be traversed on a section of the Perimeter Trail. Although *rendija* is Spanish for "cleft or crevice," this section of the canyon is comparatively wide. The area was burned in the Cerro Grande fire in 2000; however, volunteers have planted hundreds of seedlings in the canyon that are gradually restoring the habitat. There are areas of pine-oak habitat along the southern edges of the canyon, while native grasses grow on either side of the trail.

Start birding in Los Alamos at the Mitchell Trailhead near the intersection of Arizona Avenue and 45th Street. The trail at this point leads up a steep slope. You may need a walking stick or trekking poles for this part of the trail. The north portion of the Perimeter Trail crosses the Mitchell Trail. When you get to the top of the slope, the Perimeter Trail will veer to the right. Before walking along the Perimeter Trail, follow the service road toward the water tank. Approach slowly and quietly, as the trees surrounding the tank are favorite roosting spots for a large flock of Band-tailed Pigeon in season. They are quite skittish and may flush if you approach too closely.

After watching the Band-tailed Pigeons, backtrack and pick up the Perimeter Trail on your right. You can follow the Perimeter Trail to its terminus or exit on the first unmarked fork, which passes the Pajarito School and comes out at Arizona Avenue.

Target Birds

Zone-tailed Hawk It has bred in Los Alamos County and is a possibility soaring over the canyon during breeding season. All vultures should be scrutinized.

Band-tailed Pigeon The first ones arrive in early April and are present through mid- to late September.

Broad-tailed Hummingbird This high-altitude hummingbird is present from the beginning of April until mid-October.

Acorn Woodpecker Look for the flash of its white wing patch as it flies back and forth from its granary trees year-round.

American Kestrel It is present year-round and nests in the canyon.

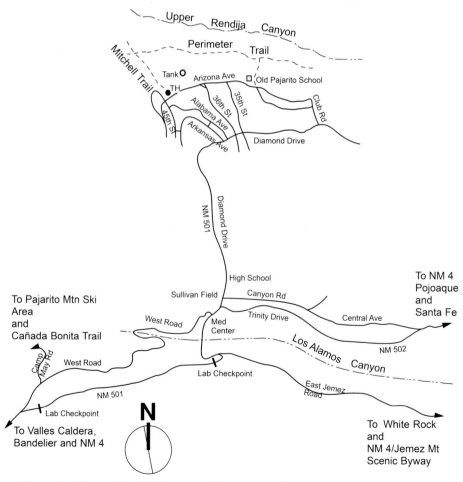

Map 9. Los Alamos, Rendija Canyon, and Perimeter Trail

Cassin's Kingbird It arrives in the canyon toward the end of April and is present through the end of August.

Violet-green Swallow It can be seen swooping over the canyon between mid-April and the end of August.

Other Birds

Species that can be seen at any time of year include Red-tailed Hawk, Eurasian Collared-Dove, White-winged Dove, Greater Roadrunner, Great Horned Owl, Downy and Hairy Woodpeckers, Northern Flicker, Pinyon and Steller's Jays, Western Scrub-Jay, American Crow, Common Raven, Mountain Chickadee, White-breasted and Pygmy Nuthatches, Canyon and Bewick's Wrens, Western Bluebird, American Robin, European Starling, Spotted and Canyon Towhees, House and Cassin's Finches (irregular), Pine Siskin, and Evening Grosbeak.

Summer species include Mourning Dove, Common Nighthawk, Common Poorwill, White-throated Swift (irregular), Western Wood-Pewee,

Upper Rendija Canyon Trail

Cordilleran Flycatcher, Say's Phoebe, Ash-throated Flycatcher, Plumbeous and Warbling Vireos, Virginia's Warbler, Spotted Towhee, Black-headed Grosbeak, Brown-headed Cowbird, and Lesser Goldfinch.

During migration, look for Rufous and Calliope Hummingbirds (mid-summer–fall), Williamson's Sapsucker (fall), Dusky Flycatcher, House Wren, Yellow-rumped Warbler, Green-tailed Towhee, Chipping Sparrow, and Western Tanager.

Wintering species include Brown Creeper, White-crowned Sparrow, and Dark-eyed Junco.

DIRECTIONS

From the intersection of I-25 (Exit 282) and St. Francis Drive (US 84/285) in Santa Fe, follow US 84/285 northwest approximately 19.5 miles to the Los Alamos exit at Pojoaque. Take NM 502 west about 18 miles to Los Alamos (NM 502 becomes Trinity Drive) to the intersection with Diamond Drive (NM 501). Turn right and travel 1.5 miles to 36th Street. Turn left and travel approximately 0.5 mile to a T intersection with Arizona Avenue. Turn left and travel about 0.3 mile to the Mitchell Trailhead on the right just before reaching 45th Street.

PARKING

There are two parking spaces at the Mitchell Trailhead; otherwise, park on the street, making sure not to block private driveways.

FEES

None

SPECIAL CONSIDERATIONS AND HAZARDS

See chapter 2 for further safety guidelines.

- Bicyclists: The Perimeter Trail is a favorite of cyclists, who often travel fast. Step off the trail when you stop to observe a bird.
- Poison ivy: This low-growing plant can be found throughout the Jemez Mountains.
- Rattlesnakes: Rattlesnakes are possible; keep your eyes and ears alert during warm weather.
- Black bears: Bears are possible in this area. Do not leave food unattended.

- Cougars: Cougars live in the Jemez Mountains and are active primarily from dusk until dawn.

- Accessibility: This trail has some challenging sections, even though most of it is fairly level.
- Restrooms: The closest restroom is in the mini-mart at the corner of Diamond Drive and Arkansas Avenue.
- Water: None available
- Picnic tables: None available
- Pets: Dogs must be kept on leash.

Dusky Grouse

CAMPING

The nearest campground is Juniper Campground at Bandelier National Monument.

GAS, FOOD, AND LODGING

The closest gas station is at the corner of Diamond Drive and Arkansas Avenue. There are numerous options for food and lodging in Los Alamos.

Pajarito Mountain Ski Area and Cañada Bonita Trail

Description

The Pajarito Mountain Ski Area, located on the eastern edge of the Jemez Mountains west of Los Alamos, is privately owned by the Los Alamos Ski Club and open to the public. The Cañada Bonita Trail (#282), a multiuse trail used by hikers, mountain bikers, and cross-country skiers as well as birders, is located across the road from the ski area and is on Santa Fe National Forest land. While the area is open year-round, it is used by birders primarily from April through October.

Start birding as you drive along Camp May Road entering the Santa Fe National Forest. The area on either side of the road was devastated during the Las Conchas fire in 2011; however, even a year later, there were signs of regrowth. Look for Northern Flickers, possibly perched on a burned snag, Western Bluebirds, and Spotted Towhees. As you approach the ski area, keep an eye out for Dusky Grouse along the road.

The first part of the Cañada Bonita Trail is smooth-packed dirt leading up a gentle grade from Camp May Road through mixed conifer woodlands starting at 9,241 feet. After 0.4 mile it emerges from the trees and hugs the edge of a hill with expansive views of Los Alamos and the distant Sangre de Cristo Mountains. There is a single-track cross-country ski trail that leads off to the left through the forest and, eventually, a meadow. The trail is very popular on crisp, clear October days when the aspens are golden. The 2-mile trail ends in a grassy meadow with summer wildflowers.

County: Los Alamos

eBird Hotspots: Pajarito Mountain and Pajarito Nordic Ski Trails website: http://www.skipajarito.com/

Target Birds

Dusky Grouse A year-round resident, the female is seen most often. From late

Cañada Bonita Trail

April through May, this bird can sometimes be seen wandering beside or on Camp May Road or the Cañada Bonita Trail. While it has been seen at all times of the day, early morning is generally the best time to spot one. The other prime time for finding one is mid-July through August, when you might encounter a hen with chicks.

Northern Goshawk This is a good location to spot this species, particularly in the summer.

American Three-toed Woodpecker It has been seen near dead trees on Pajarito Mountain and the Cañada Bonita Trail.

Clark's Nutcracker It is a year-round species that gathers nuts and seeds.

Red-breasted Nuthatch This high-elevation nuthatch, recognized by its tin horn call, is present year-round.

House Wren Look for it in low-lying shrubs from May through August.
Evening Grosbeak While this is a year-round species, the birds move around
depending on food supply (seeds, berries, caterpillars, bud worms). In this
area, they are more likely to be seen in late April and early May.

Other Birds
Species that can be seen at any time of year include Red-tailed Hawk, White-
winged Dove, Great Horned and Northern Saw-whet (calls during winter)
Owls, Downy and Hairy Woodpeckers, Steller's Jay, Western Scrub-Jay,
Common Raven, Mountain Chickadee, and White-breasted and Pygmy
Nuthatches.

Summer species include Sharp-shinned and Cooper's Hawks; Mourning
Dove; White-throated Swift; Broad-tailed Hummingbird; Red-naped Sap-
sucker; Northern Flicker; Hammond's and Cordilleran Flycatchers; War-
bling Vireo; Violet-green Swallow; Golden-crowned and Ruby-crowned
Kinglets; Hermit Thrush; American Robin; Orange-crowned, Yellow-
rumped, and Grace's Warblers; Spotted Towhee; Dark-eyed Junco; Western
Tanager; Black-headed and Pine Grosbeaks (irregular); House Finch; and
Pine Siskin.

During migration, look for Rufous Hummingbird (mid-July–September),
Olive-sided Flycatcher, Tree Swallow, Townsend's Solitaire, and Virginia's
and MacGillivray's Warblers.

DIRECTIONS
From I-25 (Exit 282) at the intersection of I-25 and St. Francis Drive (US
84/285) in Santa Fe, follow US 84/285 northwest approximately 19.5 miles
to the Los Alamos exit at Pojoaque. Take NM 502 west about 18 miles to Los
Alamos (NM 502 becomes Trinity Drive) to the intersection with Diamond
Drive (NM 501). Turn left and travel south, following NM 501 as it makes
several sharp curves toward the Los Alamos National Laboratory security
checkpoint (about 0.7 mile), where driver photo identification will be veri-
fied. After the checkpoint, continue 1.5 miles on NM 501 to Camp May
Road. Turn right and travel about 0.2 mile, where Camp May Road makes a
sharp left. Follow this road 4 miles to Pajarito Mountain Ski Area and Cañada Bonita Trailhead.

West Road is an alternative route for those wishing to avoid possible long
lines at the National Laboratory checkpoint. From Trinity Drive (NM 502)

and Diamond Drive (NM 501), travel south on Diamond Drive 500 feet and turn right on West Road. Travel on West Road along several switchbacks to the bottom of Los Alamos Canyon and back up toward the turnoff to the ski area on Camp May Road (about 2 miles).

Public transportation: During ski season only, Atomic City Transit has a bus route to the ski area with stops at many local hotels, Sullivan Field near Trinity and Diamond Drive in Los Alamos, and the Visitor Center in White Rock on NM 4 (www.AtomicCityTransit.com).

PARKING

Parking is available at the far west end of the parking lot near the base of the Aspen Lift.

FEES

None

SPECIAL CONSIDERATIONS AND HAZARDS

See chapter 2 for further safety guidelines.

- Poison ivy: This low-growing plant can be found throughout the Jemez Mountains.
- Rattlesnakes: Rattlesnakes are possible; keep your eyes and ears alert during warm weather.
- Black bears: Bears are possible in this area. Do not leave food unattended.
- Cougars: Cougars live in the Jemez Mountains and are active primarily from dusk until dawn.

FACILITIES

- Accessibility: The trail is fairly well graded with a gentle slope.
- Restrooms: Ski area and Camp May Picnic Area
- Water: Water can be purchased in the café when it is open; otherwise, bring your own.
- Picnic tables: Los Alamos County's Camp May picnic and camping areas
- Pets: Dogs must be on leash.

The nearest campground is Juniper Campground at Bandelier National Monument.

Gas, food, and lodging are available in Los Alamos. The café at Pajarito Mountain is located at the ski area and is open limited hours outside ski season.

Valles Caldera National Preserve

Description

The 89,000-acre Valles Caldera National Preserve (8,700 feet at entrance) is an area with a unique geologic history. The resulting natural history has created a varied habitat for wildlife. While the volcano that created the caldera (Spanish for "cauldron") is dormant, nearby hot springs are reminders that it is not extinct. The largest meadow grasslands area next to NM 4 is referred to as the Valle Grande. Take time to visit the exhibits in the Preserve's Valle Grande Information Center/Staging Area to learn more about the relationship between the area's geology and natural history. There are a number of different habitat types within the preserve: mixed conifer (most prevalent), wet meadow and montane grassland, ponderosa pine, and spruce-fir.

The area was a private ranch until 2000, when Congress acquired the land and created the Valles Caldera Trust to manage it. Part of the act that created the Preserve specified that it continues to be a working ranch. To protect the still-fragile habitat, most areas of the Preserve can be accessed only by reservation and on a shuttle vehicle operated by the Preserve (see website for specifics). The Preserve is constantly monitoring human impact, such as erosion. The area has been designated a Priority Important Bird Area by Audubon New Mexico. There are three areas that can be visited on your own: the road into the Staging Area, the Coyote Call Trail, and the Valle Grande Trail. There are numerous turnouts along the Jemez Mountain Byway (NM 4) that provide spectacular views of the caldera, grazing elk (fall and winter), and other wildlife, including the possibility of Bald Eagle during the winter.

Birding along the 2-mile road into the Valle Grande Staging Area offers the opportunity to see birds that reside in or migrate through montane grassland habitat. Both Eastern and Western Meadowlarks are present at this

Valle Grande

location in the summer. There is a turnout immediately after turning onto the entrance road that also allows you to look down into the area along a stream. If visiting in the summer, drive along the entrance road with your windows open to listen for the birds singing and calling and pull off at each location where the shoulder is wide enough so that other vehicles can pass. You can also bird the grassland area around the perimeter of the Staging Area.

The 3.5-mile loop Coyote Call Trail, accessed from the trailhead parking area along NM 4 near mile marker 41 (1.8 miles east of the road to the Staging Area just west of mile marker 41), leads through a montane meadow and mixed conifer habitat with related bird species. Walk up what appears to be an old road. After 40 yards the trail forks. To hike the entire trail, take the right fork. If you want a shorter hike, take the left fork and walk for a while through the ponderosa pines with views of the caldera. While rocky, the trail is fairly level in most places. Although it is not necessary to stay on the trail in this part of the Preserve, there are both brown Preserve trail markers and blue diamond cross-country ski trail markers. The Preserve requests that you sign in and out at the trailhead.

The Valle Grande Trail, just west of mile marker 44, was severely burned in a wildfire in 2011, which has impacted the species diversity; however, it remains an excellent trail for fall viewing of elk in the caldera.

County: Sandoval

eBird Hotspots: Valles Caldera Visitor Center

website: http://www.vallescaldera.gov/

Target Birds

All species listed can be seen without scheduling a trip into the backcountry.

Bald Eagle It begins arriving in early November and can most often be easily spotted from one of the turnouts along NM 4 as the eagle feeds along the East Fork of the Jemez River within the preserve.

Mountain Bluebird It nests in conifers. After breeding, it forages in the grasslands of the caldera and can be seen along the entrance road.

Vesper Sparrow It breeds in the grasslands of the caldera and can be seen both along the entrance road and at the Valle Grande Staging Area from June through September.

Savannah Sparrow A wet meadow breeder, it is not as prevalent as the Vesper Sparrow and is often overlooked during the summer season because its behavior is more secretive.

Eastern and Western Meadowlarks Both meadowlarks breed in the caldera grasslands, and this is one of the few areas in New Mexico where both species are fairly equal in abundance. Listen to the songs and calls to differentiate which species you are hearing. The best time to enjoy them is at dawn.

Brewer's Blackbird It breeds in the grasslands of the caldera and is present until early October.

Gray-crowned, Black, and Brown-capped Rosy-Finches They sometimes visit the Staging Area during the winter.

Other Birds

While the Preserve has an extensive list of birds, this section includes only those species that can be seen in non-fee areas, which are the Staging Area, Coyote Call Trail, and Valle Grande Trail.

If you visit during the winter, you might spot Green-winged Teal (in thermally heated ponds seen with a spotting scope from the easternmost turnout along NM 4).

Species that might be seen during the summer include Golden Eagle (it does not nest but forages only and is not as prevalent in public areas of the Preserve); Northern Harrier; Great Blue Heron (uncommon on river crossing along entrance road); Turkey Vulture; Sharp-shinned, Cooper's, and Swainson's Hawks; Northern Goshawk (uncommon); Killdeer; Mourning Dove; Band-tailed Pigeon (possible in forested areas); Williamson's Sapsucker; Northern Flicker; American Kestrel; Olive-sided Flycatcher (rare); Western Wood-Pewee; Cordilleran Flycatcher (forest trails); Say's Phoebe (grasslands); Warbling and Plumbeous Vireos; Violet-green Swallow; Ruby-crowned Kinglet; Western Bluebird; Hermit Thrush (forest trails); American Robin; Yellow-rumped Warbler; Green-tailed Towhee (uncommon); Chipping Sparrow; Dark-eyed Junco; Western Tanager; Red-winged Blackbird; and Pine Siskin (prevalence varies).

During migration, look for Northern Harrier (August), Lewis's Woodpecker (rare), House Wren, and Golden-crowned Kinglet (in fall on forest trails).

Species that might be seen at any time of year include Mallard (Valle Grande stock pond); Red-tailed Hawk; Wild Turkey (hunting season at the Preserve is January–March); Great Horned Owl; Hairy Woodpecker; Steller's Jay; Clark's Nutcracker; Black-billed Magpie; American Crow; Common Raven; Horned Lark; Mountain Chickadee; Red-breasted, White-breasted, and Pygmy Nuthatches; Brown Creeper; Red Crossbill (varies by year); and Evening Grosbeak.

DIRECTIONS

From the intersection of I-25 (Exit 282) and St. Francis Drive (US 84/285) in Santa Fe, follow US 84/285 northwest approximately 19.5 miles to the Los Alamos exit at Pojoaque. Take the Los Alamos Highway (NM 502) west about 12 miles to the intersection with NM 4. Turn left (exit ramp will be on the right) and travel approximately 28 miles to the entrance road to the Preserve on the right.

There is a parking lot at the Valle Grande Staging Area. There are turnouts at the entrance and along NM 4.

FEES

There is no fee for visiting the Information Center at the Staging Area; however, there may be fees for participation in various activities or visiting areas within the Preserve. See website for specifics.

SPECIAL CONSIDERATIONS AND HAZARDS

See chapter 2 for further safety guidelines.

- Winter driving: The roads can be snowpacked and icy during the winter. It is advisable to check the New Mexico Department of Transportation website (www.nmroads.com) or call the hotline (800–432–4269) for current road conditions during winter months.
- Summer monsoon season: Afternoon thunderstorms are common during the months of July and August. Bring raingear and be aware of lightning in the area.
- Black bears and cougars: Encounters with bears and cougars are possible when on the Coyote Call and Valle Grande Trails.
- Elk: The elk begin moving down from the high country toward their winter range in the caldera as the temperature cools in the fall. If you encounter elk on the Coyote Call or Valle Grande Trails, please stay at a safe distance.
- Hunting: The Preserve allows hunting, so be aware of hunter activity September through early December (elk) and late April through early May (turkey).

FACILITIES

- Accessibility: The Staging Area/Information Center is fully accessible. Inquire about accessibility of each activity. The Coyote Call and Valle Grande Trails are uneven in spots and are rated "intermediate."
- Restrooms: Staging Area
- Water: None available
- Picnic tables: Staging Area
- Pets: Pets are allowed only on the Coyote Call and Valle Grande Trails, if on leash.

CAMPING

Nearest camping to the west is at Jemez Falls or to the east at Juniper Camp-ground in Bandelier National Monument.

GAS, FOOD, AND LODGING

The closest gas, food, and lodging are available in Los Alamos and White Rock.

Along Jemez Mountain National Scenic Byway (NM 4)
Description

The route of the Jemez Mountain National Scenic Byway follows NM 4 and US 550 from near the town of White Rock on the east to San Ysidro south of Jemez Springs. Although the scenic byway continues north again on US 550 from San Ysidro, this section covers only birding sites from White Rock to the San Diego Fishing Access south of Jemez Springs on NM 4 and high-lights a number of special places to bird. Three of the stops are significant enough that a separate site description is devoted to them (Bandelier Na-tional Monument Visitor Center area, Valles Caldera National Preserve, and Jemez Falls).

There are more trails and access points along this route than are included in this section. Some that were previously good spots for birding have suf-fered fire or flood damage that severely affected the habitat and birdlife. Trails are mentioned that provide opportunities for birding across a variety of habitats.

After leaving the intersection of NM 4 and NM 502, the scenic byway travels across mesa tops with piñon-juniper habitat, passes by the Tsankawi section of Bandelier National Monument, and enters the town of White Rock. The White Rock Visitor Center on NM 4 is the pickup and drop-off point for the shuttle to the Bandelier National Monument Visitor Center, which operates between late May and mid-October when private vehicles are not permitted in the Visitor Center area of the monument between 9:00 a.m. and 3:00 p.m. (subject to change).

White Rock Overlook and Wastewater Treatment Plant: Turn left (south) on Rover Boulevard just before reaching the White Rock Visitor Center, and then left at the first street on the left (0.2 mile, Meadow Lane), and drive 0.8 mile to Overlook Road. Continue on Overlook for 0.7 mile to the end. You will pass Overlook Park and the Blue Dot Trailhead (steep, difficult trail leading down to the Rio Grande). There is a paved path to the overlook where you can gaze at a magnificent view of the Rio Grande deep in the canyon below. During the winter you might be lucky and spot a Bald Eagle. Listen for Rock and Canyon Wrens. The land on the other side of the Rio Grande is in Santa Fe County. As you leave the overlook, look for a structure on your right (0.4 mile) and pull in and park in the area to the right of the facility. You can bird the exterior of the wastewater treatment facility and the arroyo behind it. This site attracts a wide range of species, especially during migration. There is no trail, and the ground is uneven in many spots.

Return to NM 4, and travel about 8.4 miles to the entrance road to Bandelier National Monument. This road leads to the Visitor Center Area (described separately) and Juniper Campground (6,666 feet). Your entrance fee to the national monument allows you to bird in the campground. There is a large parking area along Campground Road. You can bird along the campground loops or walk along the Frey Trail, which starts from the parking lot. The initial part of the trail (0.8 mile) is relatively level and leads across the top of the mesa through piñon-juniper habitat before descending to the ruins. Of interest in the campground and mesa are several species of woodpeckers, Gray Flycatcher, Juniper Titmouse, Black-throated Gray Warbler, and Hepatic Tanager.

Six miles past the entrance to the national monument along NM 4 you will come to Bandelier National Monument's Ponderosa Group Campground (7,600 feet) on your left. Although it is a group reservation site, you can park in the parking lot while you bird. In addition to birding the perimeter of the campground, you can hike along the trail that gently descends from the campground through ponderosa forest for almost a mile to the Upper Crossing Trail. The trail leads into the Bandelier backcountry (permits required for overnight stays). This is a good location for Pygmy Nuthatch, Western Bluebird (April through September), and breeding Grace's Warbler.

Shortly after passing NM 501, which leads to Los Alamos (0.2 mile west of the Ponderosa Campground), NM 4 leaves Bandelier National Monument and enters the Valles Caldera National Preserve (described separately).

The Las Conchas Day-Use Area (8,438 feet) with nesting MacGillivray's Warbler and nearby (0.75 mile southwest) Las Conchas Trailhead are located along the east fork of the Jemez River. Both areas have suffered access complications during the summer thunderstorm season as a result of flash

Las Conchas Trail

flooding after the Las Conchas (2011) and Thompson Ridge (2013) fires. The trail leads through a canyon, with meadows and mixed conifer woodlands on the edges and slopes. American Dipper is possible along the river at any time of year but is more likely seen during the winter when there are fewer visitors. Great Horned Owl may be seen or heard in the early morning, dusk, or at night.

Traveling west on NM 4, between Las Conchas Trailhead and the turnoff to Jemez Falls, the road heads up an incline. On the far side of this hill, the east fork of the Jemez River flows parallel to the highway for a little over 2 miles, when it turns south, flowing under NM 4 shortly before the access road for Jemez Falls campground and trailhead (described separately).

The area near the junction of NM 4 and NM 126 is referred to as La Cueva and provides varied opportunities for birding.

The La Cueva Picnic Area, maintained by the Santa Fe National Forest, is popular with anglers and is another location where birders may find American Dipper. However, heavy debris-filled runoff from monsoonal rains following the fires of 2011 and 2013 has affected the stream habitat. There is a large parking area near the entrance and additional parking near the restrooms. The trail starts just past the restrooms and crosses San Antonio Creek. The willows along the creek normally harbor MacGillivray's Warbler. The trail is uneven in places. Please call the Jemez Ranger District (575–829–3535) for current conditions.

There is a nearby area worth exploring before you continue along NM 4 into the Village of Jemez Springs. Drive west along NM 126 about 0.6 mile and park in the turnout along the highway just before Twisted Juniper Road on the left. Make sure not to block the mailboxes. When you are birding in this area, please stay on the county roads and respect private property. Scan the pond for the resident Belted Kingfisher and then walk along the highway to Twisted Juniper Road. There is a marshy area along the San Antonio Creek visible from the bridge on Twisted Juniper Road that can be productive.

Return to NM 4 and head south, where San Antonio Creek joins the east fork of the Jemez River near Battleship Rock Picnic Area in the red-walled Cañon de San Diego. In addition to the opportunity to see White-throated Swifts sailing below the clifftops or perhaps a Red-tailed Hawk perched on a ledge, take the opportunity to explore some of the unique geologic features of the canyon, such as Soda Dam, where mineral deposits from underground hot springs have formed a natural travertine dam on the Jemez River. The Jemez River then continues south through the Village of Jemez Springs. The San Diego Fishing Access Area, located on a scenic bend of the river, provides an opportunity for birding in a riparian habitat at 5,800 feet.

Counties:

- Los Alamos: White Rock Overlook, Juniper Campground, Ponderosa Campground
- Sandoval: Las Conchas Trailhead, La Cueva Picnic Area, San Diego Fishing Access Area

eBird Hotspots: White Rock Overlook, White Rock Water Treatment Facility, Bandelier NM—Juniper Campground, Bandelier NM—Ponderosa Campground, Santa Fe NF—Las Conchas Trailhead, and Santa Fe NF—San Diego Fishing Access

Target Birds

Williamson's Sapsucker This sapsucker of the north can be found year-round in the Jemez Mountains, although it is not as prevalent during the winter. Look

Williamson's Sapsucker (photo by Lou Feltz)

for its sap holes in the trunks of ponderosa pine. Good locations are the Juniper Campground and Ponderosa Campgrounds (Bandelier National Monument). Note the distinct color differences between males and females.

Gray Flycatcher The sites with piñon-juniper habitat are some of the more reliable locations for this species in north-central New Mexico. Look for it between April and mid-September at the White Rock Overlook area and the Juniper Campground (Bandelier National Monument).

Pygmy Nuthatch It is a year-round species found in ponderosa pine habitat between Juniper Campground (Bandelier National Monument) and La Cueva Picnic Area.

Rock and Canyon Wrens Look for both species at the White Rock Overlook. While Canyon Wren is present year-round most years, Rock Wren can be found between April and the end of September.

American Dipper It is seen regularly at various locations along the Jemez River.

Hepatic Tanager An uncommon bird of pine-oak habitat, it is possible to see it between May and the end of September at Juniper Campground (Bandelier National Monument).

Evening Grosbeak A year-round species, it primarily travels in flocks following its food sources: berries and invertebrates.

Other Birds

The sites described in this section encompass locations with altitudes that range from 5,800 to 9,100 feet. Not all species mentioned will be found at the same time of year in the same location. For example, a species might be seen at a high elevation during the summer and then descend to a lower elevation during the winter. For specifics on a species, please refer to the annotated checklist.

Species that can be seen at any time of year at one or more locations include Mallard (rare); Scaled and Gambel's Quail (White Rock Overlook); Wild Turkey; Golden Eagle; Sharp-shinned Hawk; Eurasian Collared-Dove; White-winged Dove; Acorn, Downy, Hairy, and American Three-toed Woodpeckers; Northern Flicker; American Kestrel; Steller's and Pinyon Jays (White Rock in September); Western Scrub-Jay; Clark's Nutcracker; American Crow; Common Raven; Mountain Chickadee; Juniper Titmouse; Bushtit; Red-breasted (more common in winter at most locations), White-breasted, and Pygmy Nuthatches; Brown Creeper (more common in winter at most locations); Bewick's Wren; Golden-crowned and Ruby-crowned Kinglets; Western and Mountain Bluebirds; Townsend's Solitaire; Hermit Thrush; American Robin; Canyon Towhee; Dark-eyed Junco; and Cassin's and House Finches.

During the summer look for Great Blue Heron; Turkey Vulture; Cooper's Hawk; Mourning Dove; White-throated Swift; Black-chinned and Broad-tailed Hummingbirds; Red-naped Sapsucker; Olive-sided Flycatcher; Western Wood-Pewee; Hammond's and Cordilleran Flycatchers; Say's Phoebe; Ash-throated Flycatcher; Cassin's Kingbird; Plumbeous and Warbling Vireos; Northern Rough-winged, Violet-green, Barn, and Cliff Swallows; House Wren; Blue-gray Gnatcatcher; Northern Mockingbird (in dry years); Orange-crowned, Virginia's, Yellow-rumped, Grace's, and Black-throated Gray Warblers; Green-tailed and Spotted Towhees; Chipping and Song (Las Conchas and La Cueva) Sparrows; Western Tanager; Black-headed and Blue Grosbeaks; Brown-headed Cowbird; Pine Siskin; and Lesser Goldfinch.

During fall migration, Rufous and Calliope Hummingbirds, Cassin's Vireo, and Townsend's Warbler are possible.

In the winter you might see the following waterfowl in the Rio Grande from the White-Rock Overlook: Canada Goose, Gadwall, American Wigeon, Green-winged Teal, Bufflehead, Common Goldeneye, and Common Merganser. Other winter species include Bald Eagle, Belted Kingfisher, Peregrine Falcon (White Rock Overlook), and White-crowned Sparrow.

DIRECTIONS

From the intersection of I-25 (Exit 282) and St. Francis Drive (US 84/285) in Santa Fe, follow US 84/285 northwest approximately 19.5 miles to the Los Alamos exit at Pojoaque. Take the Los Alamos Highway (NM 502) west about 12 miles to the intersection with NM 4. Turn left (exit ramp will be on the right) and travel approximately 3.8 miles to the intersection of NM 4 and Rover Boulevard in the town of White Rock.

PARKING

- White Rock Overlook: Parking area at overlook
- Juniper Campground: Parking area near amphitheater
- Ponderosa Campground: Parking available near restrooms
- Las Conchas Trailhead: A small parking area near the trailhead and another one a short distance to the east near the rock-climbing area
- La Cueva Picnic Area: A large parking area near the entrance
- San Diego Fishing Access Area: A designated parking area

FEES

- White Rock Overlook: None
- Bandelier National Monument camping and picnic areas: Day-use fee or federal pass
- Las Conchas Day-Use Area: Day-use fee or federal pass
- Las Conchas Trailhead: None
- La Cueva Picnic Area and San Diego Fishing Access Area: Day-use fee or federal pass

SPECIAL CONSIDERATIONS AND HAZARDS

See chapter 2 for further safety guidelines.

- Winter driving conditions: The road can be snowpacked and icy during the winter. It is advisable to check the New Mexico Department of Transportation website (www.nmroads.com) or call the hotline (800–432–4269) for current road conditions before travel.

- Summer thunderstorms: Thunderstorms can develop quickly during summer afternoons. Pay attention to the sky.
- Black bears: Bears are possible in any of these areas. Do not leave food unattended.
- Cougars: Cougars are present in the Jemez Mountains. They are active primarily from dusk until dawn. If you are out at night in search of owls or start birding at dawn, be alert.
- Rattlesnakes: Rattlesnakes are possible in San Diego Canyon.
- Poison ivy and stinging nettle: Poison ivy is present at Battleship Rock, and both plants are possible throughout the Jemez Mountains.
- Mosquitoes: Mosquitoes are possible at San Diego Fishing Access Area and at a few higher locations near water.

FACILITIES

- Accessibility: Accessibility issues at each location are addressed in the description.
- Restrooms: Restrooms are available at all locations, except Las Conchas Trailhead (they are available at the nearby Las Conchas Day-Use Area).
- Water: Available only at Juniper Campground in the restroom
- Picnic tables: Non-campers can use picnic tables in unoccupied camping spaces but must vacate by 4:00 p.m. Picnic tables at the Ponderosa group site can be reserved for large groups. There are picnic tables at the Las Conchas Day-Use Area, La Cueva Picnic Area, and Battleship Rock.
- Pets: Pets are not allowed on trails in Bandelier National Monument. At any of the Santa Fe National Forest sites, pets need to be on leash at all times.

CAMPING

Bandelier National Monument: Juniper Campground, Jemez Falls, and privately owned campgrounds near Jemez Springs

GAS, FOOD, AND LODGING

Gas, food, and lodging are available in White Rock and Los Alamos at the eastern end of this route. On the western edge, food and lodging are available at Jemez Springs. Gas is available about 11 miles south of Jemez Springs

Jemez Falls

at Walatowa (Pueblo of Jemez). There are lodging and a general store in La Cueva.

Jemez Falls

Description

Located in ponderosa pine habitat, the falls are the more accessible of the two known nesting areas in New Mexico for Black Swift. The spectacular nature of the falls attracts hikers, families, and picnickers to the area, often making it challenging to watch for the elusive swifts.

Start your birding as you enter the road that leads to the campground and trailhead, keeping an eye out for accipiters zipping through the forest. Before starting down the trail, take a few moments to scan the skies above the parking area, as Black Swifts are sometimes spotted flying overhead where the view is not obscured by the forest canopy.

From the trailhead (7,957 feet) on the west (right) side of the parking lot, follow the trail through the forest as it leads gently down the hill for 0.25 mile toward the Jemez Falls Overlook. A short distance down the path, the trail to McCauley Hot Springs veers off to the right. The trail that continues to the overlook is quite rocky as it nears the falls viewing area. Informal trails leading to the falls from the overlook are not a part of the National Forest trail system. These trails are not considered safe, and their use is discouraged. Along the trail to the falls, watch and listen for nuthatches and high-altitude nesting warblers in season.

County: Sandoval

eBird Hotspots: Santa Fe NF—Jemez Falls

Target Birds

Northern Goshawk It nests in the Jemez Mountains and often can be seen hunting in this location at any time of year.

Band-tailed Pigeon Arriving in late April, this shy pigeon often can be seen perched on top of a conifer or bare snag. If few people are present, it might come in to drink in the river below the falls.

Black Swift Nesting pairs arrive in late June and are present through late August. Patience is needed to have an opportunity to see them—either flying high overhead, leaving the nest in the early morning, or returning just before dusk. They often forage a long distance from the nest and fly quite high. Unlike the White-throated Swifts or Violet-green Swallows, also in the area, the Black Swift does not circle overhead. Take time to familiarize yourself with its shape and coloring before you start birding. *Do not attempt to travel beyond the overlook viewing area in an effort to get a better view.*

White-throated Swift This is the most prevalent swift in the Jemez Mountains. It is fast flying, often zooming in the canyon above the falls. Look carefully to make sure it is not the more prevalent, but differently shaped, Violet-green Swallow.

Western Bluebird It nests in the area in cavities of the ponderosa pine.

Grace's Warbler Arriving in mid-April, it sings from and nests in the upper branches of ponderosa pine.

Red Crossbill A year-round, roving species, it follows the supply of conifer seeds/cones, often in flocks. Crossbill numbers vary greatly from year to year.

Evening Grosbeak Another year-round species that follows its food source, primarily berries and invertebrates, it travels in flocks except during breeding season, when it is secretive.

Other Birds

Species that can be seen at any time of year include Red-tailed Hawk; Williamson's Sapsucker; Downy (rare), Hairy, and American Three-toed (irregular) Woodpeckers; Steller's Jay; Clark's Nutcracker; American Crow; Common Raven; Mountain Chickadee; Juniper Titmouse; Bushtit; White-breasted and Pygmy Nuthatches; Canyon Wren; and American Dipper.

During the summer you can may find Turkey Vulture; Cooper's and Zone-tailed Hawks (irregular); Mourning Dove; Broad-tailed Hummingbird; Northern Flicker; Olive-sided Flycatcher; Western Wood-Pewee; Hammond's, Dusky, and Cordilleran Flycatchers; Plumbeous and Warbling Vireos; Violet-green Swallow; Ruby-crowned Kinglet; Mountain Bluebird (July); Hermit Thrush; American Robin; Yellow-rumped Warbler; Chipping Sparrow; Dark-eyed Junco; Western Tanager; Black-headed Grosbeak; Pine Siskin; and Lesser Goldfinch.

DIRECTIONS

From Santa Fe: At the intersection of I-25 (Exit 282) and St. Francis Drive (US 84/285), follow US 84/285 northwest approximately 19.5 miles to the Los Alamos exit at Pojoaque. Take the Los Alamos Highway (NM 502) west about 12 miles to the intersection with NM 4. Turn left (exit ramp will be on the right) and travel approximately 36 miles through the Valles Caldera to the left turn to Jemez Falls and Campground (the sign is small and easy to miss).

From Albuquerque: At the intersection of I-25 and I-40, travel north 16 miles on I-25 to Exit 242 (Bernalillo, north exit). Turn left onto US 550 and travel approximately 23.5 miles to San Ysidro. Turn right on NM 4 and travel approximately 38 miles through the Village of Jemez Springs to Jemez Falls and Campground on the right (small sign).

PARKING

Although there is a large paved parking lot near the trailhead, it is often crowded on summer weekends.

None for day use

SPECIAL CONSIDERATIONS AND HAZARDS
See chapter 2 for further safety guidelines.
- Black Swift etiquette: To assure its continued nesting success in this area, do not to attempt to encroach on its nesting area.
- Environmental fragility: Please walk on designated trails and avoid "informal" trails. Cutting live vegetation is prohibited.
- Black bears: Bears are possible in this area. Do not leave food unattended.
- Cougars: Cougars live in the Jemez Mountains and are active primarily from dusk until dawn. Be vigilant for cougars if you visit the falls at dawn or dusk to see Black Swift as they leave or return to their nests.

FACILITIES
- Accessibility: Trail is smooth as it heads down to the falls viewing area. The area near the falls is very rocky, steep, and difficult to traverse.
- Restrooms: There are restrooms near the parking area.
- Water: A water spigot is located in the campground.
- Picnic tables: There are picnic tables on the east side of the parking area.
- Pets: Pets must be on leash and under control at all times. Pick up and pack out your dog's waste.

CAMPING
Jemez Falls Campground is along the road to the trailhead.

GAS, FOOD, AND LODGING
Nearest gas is 24 miles west on NM 4 at Walatowa (Pueblo of Jemez) or 24 miles east on NM 4, then north on NM 501 at Los Alamos. Nearest lodging and general store are at La Cueva, 5 miles west on NM 4. Food and lodging are also available 14 miles west on NM 4 at Jemez Springs.

CHAPTER 7

Along the Rio Chama

General Overview

The Rio Chama, one of the tributaries of the Rio Grande, originates in the
San Juan Mountains in southern Colorado, flows through the Village of
Chama, south past Heron Lake, then into El Vado Lake before entering a
steep-walled canyon where it is designated a National Wild and Scenic River.
After emerging from the canyon, the river flows into Abiquiu Lake and then
into the Española Valley at its confluence with the Rio Grande. The Wild
and Scenic River portion, along with the adjacent plateau, is designated one
of New Mexico Audubon's Important Bird Areas.

The seven sites in this chapter start at the western edge of the Española
Valley and follow US 84 north. Highlights include two alternatives for birding the Rio Chama Wild and Scenic River; a site managed by the US Army
Corps of Engineers; two New Mexico state parks; Ghost Ranch, location of
one of artist Georgia O'Keeffe's homes; and the Village of Chama. Habitats
include Great Basin scrub, piñon-juniper, transitional/ponderosa, agricultural, and montane riparian.

Abiquiu Lake and Rio Chama Recreation Area
Description

Abiquiu (pronounced "AH-bi-kyoo") Lake (elevation 6,400 feet), a reservoir
managed by the US Army Corps of Engineers, lies along the Rio Chama in a
valley between the Jemez Mountains to the southwest and a group of mountain ridges that make up the southernmost spur of the San Juan Mountains
to the north. The Cerro Pedernal in the Jemez, made famous in paintings by
Georgia O'Keeffe, looms nearby. The Rio Chama Recreation Area is located
400 feet below the tall earthen dam where the river flows out of the lake and

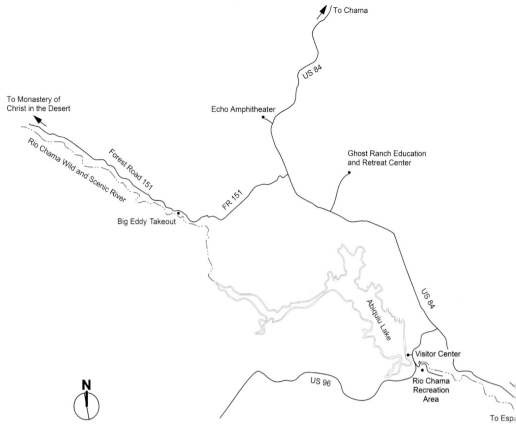

Map 10. Abiquiu Lake, Ghost Ranch, and Rio Chama Wild Rivers

downstream from Española, where it joins the Rio Grande. The area surrounding the Visitor Center and campgrounds is in piñon-juniper habitat, while the Rio Chama Recreation Area has a narrow band of riparian habitat at the base of towering reddish sandstone cliffs.

Start birding in the Village of Abiquiu as you drive northwest through agricultural areas along the cottonwood-lined US 84. Black-billed Magpie can often be seen in the fields, and Lewis's Woodpecker is frequently spotted as it flies between the cottonwood trees. This section of US 84 winds through private property where the road shoulders often are not wide enough to pull over. During spring and fall migration small to large flocks of gulls, ibis, and other shorebirds may be seen if the agricultural fields are under irrigation. Be sure not to obstruct private drives if you stop for birding. As the road curves up the hill, look for a turnout on your left. Not only does it provide views of the Española Valley—especially spectacular in October when the cottonwood gallery has turned golden—but it also allows you to search the cliffs on the northeast side of the highway for whitewash (area covered with bird droppings) and perhaps the opportunity to spot a Peregrine Falcon or Golden Eagle soaring above the cliff tops.

Begin your visit at the Abiquiu Lake Visitor Center (open 8:00 a.m. to 4:00 p.m. Monday through Friday; closed holidays and weekends), located 1.5 miles off US 84 on NM 96. You can pick up a map and learn more about the area from the displays. From the Visitor Center, you can walk along all or part of the Piñon Trail for piñon-juniper species, scanning for waterfowl and raptors.

As you exit the Visitor Center parking lot, cross the road turning right. Open sunrise to sunset, the road zigzags down the earthen dam to the power plant and the Rio Chama Recreation Area. At the base of the dam, turn left into the Rio Chama Recreation Area. Bird along the river, check the scrub vegetation on the hillside behind the restroom, and study the cliffs across the river. The area at the base of the cliff above the Recreation Area has water seepage and vegetation that attracts birds, especially during migration.

Red-tailed Hawks nest on the high cliffs overlooking the river. Occasionally Peregrine Falcons and Golden Eagles will cruise through this part of the canyon. During spring and summer, look for Cliff Swallows flying in and out of their mud nests located high on the canyon walls.

If you wish to explore the Rio Chama farther downstream, leave your vehicle in the parking lot and walk along the dirt road, part of the northern branch of the Old Spanish Trail, which served as a trade route between northern New Mexico and Spanish settlements in California in the early 1800s. The Army Corps of Engineers' property ends a short distance down the road, designated by a fence across the river. The road beyond this point is not maintained, is rutted, and has few turnaround locations.

County: Rio Arriba

eBird Hotspots: Abiquiu Lake—Visitor Center and Overlook and Rio Chama Recreation Area

Target Birds

Common Merganser It is possible to spot it on the lake; however, it is more likely seen on the river below the dam at any time of year.

Rio Chama Recreation Area

Double-crested Cormorant It fishes on the lake and along the river between March and September, and breeding colonies have been documented.
Bald Eagle More than 15 Bald Eagles have been counted on the Army Corp of Engineers Annual Eagle Survey at the lake in early January.
Belted Kingfisher It can be found perched along the river at any time of year.
Loggerhead Shrike Look for it on fence posts and signs along the highways near or within the Abiquiu Lake Recreation Area.
Canyon and Rock Wrens Canyon Wren, present year-round, is found along the cliffs at the Rio Chama Recreation Area. Rock Wren, a summer resident, often found in the rock-armored slope of the dam, can be found at both locations.
American Dipper Although not common, it has been spotted along the river below the dam at all times of year.

Other Birds
Species that can be seen at any time of year include Canada Goose, Mallard, Pied-billed Grebe, Great Blue Heron, Red-tailed Hawk, American Coot, Pinyon Jay, Western Scrub-Jay, Common Raven, Bushtit, Juniper Titmouse, and Canyon Towhee.

In the summer, look for Turkey Vulture; Mourning Dove; Spotted Sandpiper; Broad-tailed Hummingbird; Black and Say's Phoebes; Ash-throated Flycatcher; Cassin's Kingbird; Violet-green, Barn, and Cliff Swallows; Western Bluebird; Northern Mockingbird; Yellow Warbler; Lark Sparrow; Bullock's Oriole; Brewer's Blackbird; Brown-headed Cowbird; House Finch; and Pine Siskin.

During migration you might see Western Grebe, Rufous Hummingbird, and Virginia's and MacGillivray's Warblers.

In the winter it is possible to see Gadwall, Bufflehead, Common and Barrow's (irregular) Goldeneye, American Robin, Song and White-crowned Sparrows, Dark-eyed Junco, and Mountain Bluebird.

DIRECTIONS

From Santa Fe: At the intersection of I-25 (Exit 282) and St. Francis Drive (US 84/285), follow US 84 northwest approximately 57 miles (through Española and Abiquiu) to NM 96. Turn left and travel 1.5 miles to the Visitor Center.

From the Village of Abiquiu: Travel 6.5 miles northwest on US 84 and turn left onto NM 96 toward the Visitor Center.

PARKING

There is a large parking area around the Visitor Center and at the Rio Chama Recreation Area.

FEES

There is no fee at the Visitor Center or at the Rio Chama Recreation Area. There are fees for camping and boat ramp use.

SPECIAL CONSIDERATIONS AND HAZARDS

See chapter 2 for further safety guidelines.

- Winter driving conditions: US 84 can be snowpacked and icy during the winter. It is advisable to check the New Mexico Department of Transportation website (www.nmroads.com) or call the hotline (800–432–4269) for current road conditions during winter months.
- Rattlesnakes: Rattlesnakes are possible. Keep your eyes and ears alert during warm weather. See chapter 2 for additional safety information.

FACILITIES

- Accessibility: The Visitor Center area is level and accessible. The trail near the Overlook is uneven and leads over a rock outcropping. The Rio Chama Recreation Area is level.
- Restrooms: Behind the Visitor Center at the Overlook and Rio Chama Recreation Area
- Water: Visitor Center and Overlook and Riana Campground
- Picnic tables: Visitor Center and Overlook and Rio Chama Recreation Area
- Pets: Pets must be on leash.

CAMPING

Camping is available at the Riana Campground.

GAS, FOOD, AND LODGING

A gas station, convenience store, restaurant, and lodging are available in Abiquiu.

Ghost Ranch Education and Retreat Center and Echo Amphitheater
Description

Situated in the mouth of a canyon (6,420 feet) at the foot of bluffs and mesas striated in shades of russet and alabaster, the current Ghost Ranch Education and Retreat Center was originally a working ranch, Rancho de los Brujos (Spanish for "witches"), and later a dude ranch, near where painter Georgia O'Keeffe purchased her first home in New Mexico. There are many stories about how it came to be called Ghost Ranch. During the nineteenth century, cattle rustlers used the canyon to hide cattle. The cottonwood tree in front of the present library is said to have been the site where one of the rustlers was hanged. Later residents have claimed to have heard his ghost.

The area also has rich archaeological and paleontological histories. Artifacts continue to be discovered from early hunter-gatherer peoples who lived in the area as early as 6,000 years ago. During the Triassic period, dinosaurs roamed here. Their fossils have been excavated from two quarries: the currently active Hayden (site of the oldest whole-body dinosaur fossil in North America) and *Coelophysis* (named by the New Mexico state legislature as the state dinosaur and designated by the National Park Service as a National Natural Landmark). You can visit the two museums (in the same building) devoted to these topics (open 9:00 a.m. to 5:00 p.m., Monday through Saturday, and 1:00 p.m. to 5:00 p.m. on Sunday).

The property encompasses several habitat types: Great Basin scrub, piñon-juniper, and a stream with montane riparian habitat in Box Canyon. The Retreat Center welcomes birders, hikers, and photographers who visit the ranch to enjoy its natural history and scenic beauty. It also offers classes each spring on the birds of Ghost Ranch (check for dates at ghostranch. org).

Start your birding as you drive along the 1-mile, well-maintained dirt road into the main part of the ranch. At any time of year Western Scrub-Jay normally can be encountered, as well as Western or Mountain Bluebirds.

Take the left fork in the road, drive past the museums, and park in front of the Welcome Center (open year-round 8:00 a.m. to 5:00 p.m., Monday through Saturday, and 9:00 a.m. to 5:00 p.m. on Sunday) to pick up a trail map and pay the day-use fee.

During spring and summer the large cottonwoods near the Welcome Center are host to Downy and Hairy Woodpeckers, Northern Flicker, Cassin's and Western Kingbirds, Western Bluebird, American Robin, White-breasted Nuthatch, Yellow Warbler, Bullock's Oriole, Black-headed Grosbeak, and Lesser and American Goldfinches. Barn Swallows nest in the eaves of the adobe buildings. During summer, take time to investigate the alfalfa field for Lark Sparrow and Brown-headed Cowbird.

The Box Canyon Trail provides a good opportunity for birding. Start your hike from the two public parking areas at the Welcome Center or near the museums. The dirt road to Box Canyon leading from the Arts Center and Long House starts out in scrub habitat. The trail forks but is well signed. It hugs the base of a mesa and passes a cattail-bordered pond. Although usually not saturated with birds, during the spring you may encounter Yellow Warbler, Yellow-breasted Chat, and Common Yellowthroat in the cattails, and Spotted Towhee and Blue and Black-headed Grosbeaks singing from the nearby cottonwoods. It is easy to see several species of swallows feeding over this pond as well as White-throated Swift flying overhead. The trail crosses a creek, Arroyo del Yeso, and becomes more uneven as it ventures farther into the canyon. The riparian habitat on either side of the trail is often productive. Continue to follow the signs up the canyon. Where the trail reaches a small pool, it becomes more difficult and requires climbing over some large rocks. Many birders return at this point.

Trail to Box Canyon

To visit the Echo Amphitheater, part of Carson National Forest, drive back to US 84, turn right, and travel 3 miles north. The turnoff is on your left. The naturally formed amphitheater is located in piñon-juniper habitat. There is a gradually sloping paved nature trail (#46) leading from the parking lot, past picnic tables, to the base of the amphitheater. While searching for Cliff Swallow nests, look closely at the lines of stratification in the sandstone—they outline ancient sand dunes. As the name implies, voices and sounds can be heard as echoes. The Day-Use Area is open from 7:00 a.m. to 7:00 p.m.

County: Rio Arriba

eBird Hotspots: Ghost Ranch and Carson NF—Echo Amphitheater

Target Birds

White-throated Swift It can be seen zooming below the tops of the mesas at both locations from late March through September.

Ash-throated Flycatcher This crested flycatcher of dry terrain can be heard calling what sounds like a harsh *come here* between mid-April and the end of August.

Say's Phoebe Unlike other phoebes, it prefers open, barren areas. Look for it perched on fence posts and scrub or hovering "helicopter-like" from the end of March through September.

Pinyon Jay While this roving species is not guaranteed, flocks of these all-blue, shorter-tailed jays are often first detected by their wicked-sounding call.

Juniper Titmouse Listen for its quiet tapping in piñon pine or juniper trees in Box Canyon and the Echo Amphitheater year-round.

Bewick's Wren It can be found year-round in scrubby woods, such as along the Box Canyon Trail.

Western and Mountain Bluebirds Both bluebirds can be found at the ranch year-round. Western Bluebird nests in the cottonwoods near the Welcome Center. Mountain Bluebird nests at forest edges and is not as easily seen. After nesting, these birds prefer open areas.

Western Tanager This striking tanager arrives in late April or early May and can be seen through early September.

Other Birds

Species that can be seen at any time of year include Mallard, Northern Harrier, Cooper's and Red-tailed Hawks, Rock Pigeon, Eurasian Collared-Dove, White-winged and Mourning Doves, Great Horned Owl, Downy and Hairy Woodpeckers, Northern Flicker, Peregrine Falcon, Western Scrub-Jay, Black-billed Magpie, American Crow, Common Raven, Mountain Chickadee, White-breasted Nuthatch, American Robin, Canyon Towhee, House Finch, Evening Grosbeak, and House Sparrow.

In summer look for Turkey Vulture; Killdeer; Black-chinned and Broad-tailed Hummingbirds; American Kestrel; Western Wood-Pewee; Cassin's and Western Kingbirds; Plumbeous and Warbling Vireos; Violet-green, Barn, and Cliff Swallows; House Wren; Blue-gray Gnatcatcher; Northern Mockingbird; Virginia's, Yellow, and Black-throated Gray Warblers; Chipping and Lark Sparrows; Black-headed and Blue Grosbeaks; Indigo Bunting (irregular); Red-winged and Brewer's Blackbirds; Western Meadowlark; Brown-headed Cowbird; Bullock's Oriole; and Lesser and American Goldfinches.

During migration you might spot Rufous Hummingbird (late summer through fall); Williamson's and Red-naped Sapsuckers; Tree Swallow; Ruby-crowned Kinglet; Hermit Thrush; Orange-crowned, MacGillivray's, Yellow-rumped, and Townsend's (fall) Warblers; Green-tailed Towhee; Brewer's and

Vesper Sparrows (fall); Lincoln's Sparrow; Lazuli Bunting; Pine Siskin; and Evening Grosbeak.

Species that are normally present during the winter include Townsend's Solitaire, White-crowned Sparrow, Dark-eyed Junco, and Cassin's Finch.

DIRECTIONS

From the intersection of I-25 (Exit 282) and St. Francis Drive (US 84/285) in Santa Fe, drive northwest on US 84 approximately 64 miles (through the towns of Española and Abiquiu) to the turnoff to Ghost Ranch Education and Retreat Center (turnoff is approximately 13 miles northwest of Abiquiu and 6 miles beyond the turnoff to Abiquiu Lake). Turn right and take the entrance road 1 mile to the Welcome Center.

The left turn to Echo Amphitheater is 3 miles beyond Ghost Ranch on US 84. The small sign to this picnic area is easy to miss.

PARKING

Public parking at Ghost Ranch is available at the Welcome Center and at the museums. There is a large parking lot at Echo Amphitheater.

FEES

Ghost Ranch: $3.00 per person if not attending a program

Echo Amphitheater: Day-use fee of $2.00 per car, $1.00 per individual walk-in, or federal pass

SPECIAL CONSIDERATIONS AND HAZARDS

See chapter 2 for further safety guidelines.

- Thunderstorms: Afternoon thunderstorms can form quickly during the summer monsoon season, and flash flooding is possible in Box Canyon.
- Winter driving conditions: US 84 can be snowpacked and icy during the winter. It is advisable to check the New Mexico Department of Transportation website (www.nmroads.com) or call the hotline (800–432–4269) for current road conditions during winter months.
- Rattlesnakes: Rattlesnakes are possible. Keep your eyes and ears alert during warm weather.
- Cougars: Cougars have been spotted in the canyons surrounding Ghost Ranch.

- Archaeological sensitivity: This is a site of geologic and cultural significance. Please stay on the designated trails. Collecting plants, rocks, fossil fragments, or artifacts is prohibited.

FACILITIES

- Accessibility: The area around the main buildings and fields is level. Box Canyon Trail is level for 0.3 mile and then becomes narrow, uneven, and rocky. The Echo Amphitheater Nature Trail is paved.
- Restrooms: Welcome Center at Ghost Ranch and the parking lot at Echo Amphitheater
- Water: Water can be purchased in the Welcome Center at Ghost Ranch and is available at Echo Amphitheater.
- Picnic tables: There are picnic tables outside the dining hall at Ghost Ranch; several picnic tables are along the Nature Trail at Echo Amphitheater.
- Pets: Pets must be on leash.

CAMPING

The campground at Ghost Ranch can be reserved ahead of time. Camping is available at Echo Amphitheater or at Abiquiu Lake Recreation Area.

GAS, FOOD, AND LODGING

The closest gas station is in Abiquiu. Snacks can be purchased in the trading post and coffee shop located in the Ghost Ranch Welcome Center; there is a cafeteria in the dining hall. Rooms and casitas can be reserved at Ghost Ranch; other lodging is available in Abiquiu.

Rio Chama Wild and Scenic River

Description

Three alternatives are provided for birding along the segment of the Rio Chama designated a Wild and Scenic River by Congress in 1988: on a float trip along the river from El Vado Ranch (Cooper's) to Big Eddy, on a float trip from just south of the Monastery of Christ in the Desert to Big Eddy, or by vehicle via Forest Road 151, which winds along the river from US 84 to the monastery.

The Rio Chama in this area flows through a canyon flanked by majestic sandstone bluffs whose layers represent rocks of the Triassic, Jurassic, and

Cretaceous geologic periods. Those who bird by river trip have an opportunity to hike into a canyon to see Jurassic dinosaur tracks. Six miles of the river are bounded by designated wilderness area. The land on either side of the river is managed by an array of state, tribal, and federal agencies, including the Bureau of Land Management, US Forest Service, Jicarilla Apache Nation, and New Mexico Department of Game and Fish, as well as some privately owned lands. The area also has been designated by New Mexico Audubon as an Important Bird Area, as it provides habitat for nesting raptors and five species of state conservation concern, including Dusky Flycatcher, Plumbeous Vireo, Western Bluebird, Pygmy Nuthatch, and Grace's Warbler.

Birding by river trip (either from El Vado Ranch, 3-day trip; or FR 151, day trip; see site descriptions and website for more detailed information and options): As you travel down the river, you pass through wide areas of the canyon as well as sections where it is narrow with sheer cliffs towering above. Willows and cottonwoods line the riverbanks at most spots. Ponderosa pine and fir grow all the way down to the river in some locations, especially north-facing slopes, while other sections are more barren with piñon-juniper habitat. The diverse habitats provide opportunities for a wide variety of avian species. Water levels reflect releases from the upstream El Vado Lake Dam. While most segments of the river are rated Class I, calm and easy flowing, which provide better opportunities for observing birds, there are locations with Class II and III rapids.

Birding along FR 151: Start your birding as you exit US 84. Until the road reaches the river at Big Eddy, it travels through Great Basin shrub and grasslands that provide habitat for migrating sparrows and other passerines. After about 3 miles, the road enters Santa Fe National Forest, makes a bend, and descends into piñon-juniper habitat with rocky outcroppings. Look for Rock Wren in this location. The first stop along the river is the Big Eddy Takeout. Spend time birding along the river, and check the cliffs for White-throated Swifts and nesting Violet-green Swallows. Just beyond is the Whirlpool Road Access Camp with primarily piñon-juniper habitat. After Whirlpool, the road hugs the edge of the cliff and then dips down briefly through a stand of cottonwoods. The canyon then opens up. You will come to an old bridge where the Continental Divide Trail crosses the river. Two miles farther along is Oak Point Road Access Camp (no facilities), which has dense vegetation and an opportunity for birding. You also can bird along

Rio Chama

the loop road of the Rio Chama Campground a little more than a mile farther down the road. FR 151 ends at the boundary of the Monastery of
Christ in the Desert and Chavez Canyon River Access, the launch location
for starting a float trip ending at Big Eddy.

County: Rio Arriba

eBird Hotspots: Rio Chama—Big Eddy Take Out

website: http://www.blm.gov/nm/st/en/prog/recreation/taos/rio_chama_
wsr.html

Target Birds

Peregrine Falcon The cliffs and river provide superb habitat for this falcon,
which is frequently seen on float trips.

Spotted Sandpiper It nests along the river and can be seen most easily from a
raft or kayak. Also search for it teetering on rocks from day-use or camping
areas.

Plumbeous Vireo Look for it in both ponderosa pine and piñon-juniper habitats along the river between May and early September.

Violet-green and Cliff Swallows Both swallows nest in the cliff faces along the river, arriving by mid-April. Cliff Swallows leave first and are generally gone by early August, while Violet-green Swallows can be seen into early September. Swallows do not associate with their nests late in the season.

Rock and Canyon Wrens Canyon Wren is more easily heard than seen. Listen for its melodic, descending song during breeding season to try to locate it on ledges of cliffs. Rock Wren is not as elusive as the Canyon Wren. Look for it bobbing on outcroppings or rubble areas on the cliffs.

American Dipper It nests near the launch area at El Vado Ranch. Dippers can be seen all along the river, particularly in swiftly flowing rocky sections.

Virginia's Warbler Look for this drab warbler on the ground or low in branches of piñon-juniper or ponderosa pine. It can be seen from May through mid-September.

Grace's Warbler It prefers the high branches of ponderosa pines found along the middle section of the river run, often flitting deep within the treetops. Listen for its song to help you locate it.

Other Birds

Along the river from El Vado Ranch to the monastery (May 1 to Labor Day):

The birds will vary depending on the habitat along the river. Species present all summer include Canada Goose; Mallard; Common Merganser; Wild Turkey (in the canyons above the river); Great Blue Heron; Turkey Vulture; Osprey (irregular); Golden Eagle; Cooper's (irregular) and Red-tailed Hawks; White-throated Swift (in areas adjacent to tall cliffs); Broad-tailed Hummingbird; Hairy Woodpecker; Northern Flicker; American Kestrel; Western Wood-Pewee; Black and Say's Phoebes; Ash-throated Fly-catcher; Cassin's Kingbird; Pinyon and Steller's Jays; Common Raven; Warbling Vireo; Northern Rough-winged Swallow; Mountain Chickadee; Red-breasted and White-breasted Nuthatches; Western Bluebird; American Robin; Yellow, Yellow-rumped, and Black-throated Gray Warblers; Spotted Towhee; Chipping Sparrow; Hepatic (irregular) and Western Tanagers; Black-headed and Blue Grosbeaks; Indigo Bunting; Red-winged and Brewer's Blackbirds; and Common Grackle.

At night listen for Flammulated and Great Horned Owls, Western Screech-Owl, and Northern Saw-whet Owl.

During migration it is possible to spot Cinnamon Teal, Common Night-hawk, Hammond's Flycatcher, Rufous Hummingbird (fall), Tree Swallow, House Wren, MacGillivray's Warbler, Green-tailed Towhee, and Lazuli Bunting.

Along the river from the monastery to Big Eddy (on the river or along FR 151):

Species that can be seen at any time of year include Mallard, Golden Eagle, Pinyon Jay, Western Scrub-Jay, Black-billed Magpie, Clark's Nutcracker, Common Raven, European Starling, and Spotted Towhee.

During the summer you may observe Mourning Dove, White-throated Swift, Broad-tailed Hummingbird, Western Wood-Pewee, Ash-throated Flycatcher, Northern Rough-winged and Barn Swallows, Yellow Warbler, Yellow-breasted Chat, Black-headed and Blue Grosbeaks, Indigo Bunting, Brewer's Blackbird, and Brown-headed Cowbird.

During migration, look for Rufous Hummingbird (fall), House Wren, Blue-gray Gnatcatcher, Chipping and Lark Sparrows, and Lazuli Bunting.

In the winter you may find Townsend's Solitaire, White-crowned Sparrow, and Dark-eyed Junco.

Along FR 151, Big Eddy to US 84:

Species that might be seen at any time of year include Red-tailed Hawk, American Kestrel, Pinyon Jay (piñon-juniper), Common Raven, and Horned Lark (grasslands).

During the summer, look for Say's Phoebe, Barn Swallow, Lark Sparrow, and House Finch.

During migration, you might spot Sage Thrasher, Blue-gray Gnatcatcher, and Chipping, Brewer's, and Vesper Sparrows.

DIRECTIONS

From Santa Fe to Forest Road 151: From the intersection of I-25 (Exit 282) and St. Francis Drive (US 84/285), follow US 84 north approximately 66 miles (through Española and Abiquiu) to FR 151 (about 2.5 miles past the right turn to Ghost Ranch Education and Retreat Center). FR 151 continues toward the Rio Chama Wild and Scenic River and to Big Eddy, Chavez Canyon River Access, and the Monastery of Christ in the Desert.

From Santa Fe to El Vado Ranch (Cooper's): From the intersection of I-25 (Exit 282) and St. Francis Drive (US 84/285), follow US 84 north approximately 96 miles (through Española, Abiquiu, and Tierra Amarilla) to

NM 112. Turn left onto NM 112 and travel about 14 miles to the left turn to El Vado Ranch and launch site.

From Chama to El Vado Ranch (Cooper's): From the junction of US 84/64 and NM 17 (south of the Village of Chama), travel south on US 84 approximately 11.5 miles to NM 112. Turn right onto NM 112 and travel about 14 miles to the left turn to El Vado Ranch and launch site.

From Chama to FR 151: From the junction of US 84/64 and NM 17 (south of the Village of Chama), travel south on US 84 about 41.5 miles to FR 151. Turn right and travel about 5 miles to the river.

PARKING

For those taking the river trip, there is parking at El Vado Ranch. For those driving along FR 151, there is parking at Big Eddy and Chavez Canyon River Access. You can park near the restrooms at Rio Chama Campground. Although there is no designated parking, there are suitable locations in the primitive camping areas.

FEES

None for day use or camping along FR 151. Permits, assigned by lottery in February, are required for individuals floating in the upper canyon through the wilderness from May 1 through Labor Day weekend. These are obtained from the BLM Taos Field Office (575–758–8851). There is a fee for application and launch. Group size is limited to 16 people. No advance permits are required, and no fee is charged for launching on the lower section of the canyon below the monastery at Chavez Canyon River Access. There is a $5.00 day-use fee for use of the launch site at El Vado Ranch. For those traveling with a professional river outfitter, fees are included in the price.

SPECIAL CONSIDERATIONS AND HAZARDS

See chapter 2 for further safety guidelines.

- Road conditions: FR 151 in many areas is one lane with turnouts; vehicles can appear "out of nowhere." Drive slowly on blind curves, especially the ones with a cliff edge. Travel can be difficult during or following rain. Snow is possible in winter.
- Rattlesnakes: Campground signs warn of rattlesnakes. Keep your eyes and ears alert.
- Black bears and cougars: Be prepared for these mammals, especially

while primitive camping along the river, where it can be difficult to store food securely.

- Insects: There are mosquitoes along the river during the summer and yellow jackets from late August to mid-September.
- Cell phone service: There is no cell phone coverage in this area.
- River safety: Review the BLM safety guidelines at http://www.blm.gov/nm/st/en/prog/recreation/recreation_activities/boating_and_kayaking.html#nmriversafety.

FACILITIES

- Accessibility: The dirt roads in the primitive camping areas along FR 151 are fairly level. Some of the float trip camping sites along the river can require walking up a steep riverbank.
- Restrooms: For those taking a river trip, there are restrooms at El Vado Ranch. There are no facilities on the river itself. Solid human waste must be carried out in a washable, reusable container; no plastic bags are allowed. Along FR 151, there are restrooms at Big Eddy, Whirlpool Road Access Camp, Rio Chama Campground, and the Chavez Canyon River Access.
- Water: Water is available only at El Vado Ranch. Travelers must bring their own.
- Picnic tables: At only one of the campsites in the Rio Chama Campground along FR 151
- Pets: Along FR 151, pets must be leashed and under control at all times, and their waste must be picked up and bagged.

CAMPING

There are designated primitive camping sites along the river. Rio Chama Campground, maintained by the Santa Fe National Forest, is along FR 151. There are campgrounds at El Vado State Park adjacent to El Vado Ranch.

GAS, FOOD, AND LODGING

El Vado Ranch has cabins and limited food at its fishing supply store. The closest gas, restaurants, and motels are available in Chama. The turnoff to FR 151 is approximately 15 miles north of the town of Abiquiu, where there are a gas station, convenience store, restaurant, and lodging.

El Vado Lake

El Vado Lake State Park

Description

Known for its nesting Osprey, El Vado Lake (6,900 feet) sits in a valley formed by the reservoir along the Rio Chama, with ponderosa pine forest habitat surrounding it and sandstone cliffs to the west. The campground areas on the east side of the lake are situated in a grassland habitat. The access road crosses the Rio Chama Wildlife Management Area through piñon-juniper woodlands tucked among large boulders. The lake is fed from the north by the Rio Chama, originating in the San Juan Mountains of Colorado. The Rio Chama continues south below the lake. During drought years, water is released to meet farmers' water needs downstream, which may leave very little water in the lake. When this happens, some boat ramps are closed, which reduces noise and activity on the lake.

The Rio Chama Trail (5.5 miles one way) starts at Shale Point, climbs up to a mesa, and then drops down, follows the canyon, and ends at Heron

Lake State Park. The first part of the trail is fairly easy with a gentle grade. Because it crosses the Rio Chama Wildlife Management Area, it is open only from Memorial Day weekend through November 15.

In the winter, ice fishing and cross-country skiing are popular.

Start your birding as you drive along NM 112 after turning off US 84. The highway passes through farms and grasslands. Check the power poles for American Kestrel and Loggerhead Shrike and the fence lines for sparrows. After 12.2 miles, turn right onto the access road into the state park.

County: Rio Arriba

eBird Hotspots: El Vado Lake State Park

website: http://www.emnrd.state.nm.us/SPD/elvadolakestatepark.html

Target Birds

Common Merganser When there is sufficient water, it nests on El Vado Lake and can be seen year-round.

Western Grebe It is generally present from March through November.

Bald Eagle One or more winter near the lake and often can be seen near the lake.

Osprey Osprey platforms, a highlight of El Vado Lake, attract several nesting pairs each year. Ospreys arrive by early April and are present through early September.

American Kestrel It is a year-round species that hunts in open areas.

Mountain Bluebird A cavity nester, it prefers forest edges during its breeding season, moving to open grasslands for fall and winter foraging.

Chipping Sparrow Identified by its harsh, trilling song, it returns to the area to breed as early as April and may remain as late as October.

Other Birds

Species that can be seen at all times of year include Canada Goose; Mallard; Great Blue Heron; Sharp-shinned, Cooper's, and Red-tailed Hawks; Ring-billed Gull (irregular); Northern Flicker; Black Phoebe; Pinyon Jay (irregular); Western Scrub-Jay; Black-billed Magpie; American Crow; Common Raven; Mountain Chickadee; White-breasted Nuthatch; Western Bluebird; American Robin; European Starling; Red-winged Blackbird; and House Finch.

During the summer, look for Turkey Vulture, Spotted Sandpiper, Black-chinned and Broad-tailed Hummingbirds, Western Wood-Pewee, Say's Phoebe, Plumbeous Vireo, Violet-green and Cliff Swallows, Yellow Warbler (along the Rio Chama Trail), Spotted Towhee, Black-headed Grosbeak (Rio Chama Trail), Western Meadowlark, Brewer's Blackbird, Brown-headed Cowbird, and Lesser Goldfinch.

During migration it is possible to spot Green-winged Teal, Eared Grebe, Forster's Tern, Red-naped Sapsucker, Tree Swallow, Blue-gray Gnatcatcher, Virginia's Warbler, Green-tailed Towhee, and Lark Sparrow.

DIRECTIONS

From Santa Fe: From the intersection of I-25 (Exit 282) and St. Francis Drive (US 84/285), follow US 84 north approximately 96 miles (through the towns of Española and Tierra Amarilla toward Chama) to NM 112. Turn left and begin birding along NM 112. The El Vado Lake State Park access road is approximately 12.2 miles from the turnoff from US 84.

From Chama: From the intersection of US 84/64 and NM 17, 1.5 miles south of town, travel south on US 84 approximately 11.5 miles to NM 112. Turn right and travel approximately 12.2 miles to the park.

PARKING

Parking is available at the park office, the El Vado Dam Day-Use Area near the dam, and several other locations throughout the park.

FEES

Day-use fee or state park annual pass. A GAIN Permit (see chapter 2) is required to hike on the Rio Chama Trail.

SPECIAL CONSIDERATIONS AND HAZARDS

See chapter 2 for further safety guidelines.
- Rattlesnakes: Rattlesnakes are possible here.
- Yellow jackets: These stinging insects are often in the camping/picnic areas in late summer.
- Winter driving conditions: US 84 and NM 112 can be snowpacked and icy during the winter. It is advisable to check the New Mexico Department of Transportation website (www.nmroads.com) or call the hotline (800–432–4269) for current road conditions during winter months.

FACILITIES

- Accessibility: The state park area is level and has accessible tables, grills, and restrooms. The Rio Chama Trail is strenuous, often rocky, and uneven.
- Restrooms: Several restrooms are located throughout the park.
- Water: Available in campgrounds
- Picnic tables: El Vado Dam Day-Use Area
- Pets: Park rules require that dogs be kept on leash.

CAMPING

There are several camping areas at El Vado State Park.

GAS, FOOD, AND LODGING

The closest gas and food are available in Chama. Cabins and limited supplies are available at El Vado Ranch along the Rio Chama on NM 112.

Heron Lake State Park

Description

Set primarily in a transitional ponderosa pine habitat at 7,100 feet, Heron Lake has several breeding pairs of Osprey each summer. The area is an ecotone fringing on piñon-juniper habitat, thus creating a broader range of potential bird species.

The reservoir was created on Willow Creek near its confluence with the Rio Chama as a storage area for the San Juan River Project, which diverts water from tributaries of the San Juan and Colorado Rivers on the western side of the Continental Divide. The water is pumped through the Azotea Tunnel west of the Village of Chama and then into Heron Lake. The outflow joins the Rio Chama just below the dam. The area where the two courses of water come together can be seen from the Rio Chama Trail. The water diversion, which provides water to the central Rio Grande valley, means the lake levels are affected by the amount of winter snowpack. In drought years, the water levels are low, and in wet years, the lake can hold as much as 399,980 acre-feet.

Begin your birding along NM 95 as you drive through the farmlands of the Los Ojos area. This area hosts species that prefer grassland and agricultural habitats.

Stop at the pay station and the Visitor Center, if open, to pick up a map and pocket trail guides. This is a good area to park and explore along the East Meadow Trail (2.4 miles one way). The trail leading across an open meadow has a firm surface and is generally flat, with a few hilly areas. There is a trail map and guide online at http://www.emnrd.state.nm.us/SPD/documents/EastMeadowTrailGuidereduced.pdf.

In order to bird the woodland areas of the park, walk west from the Visitor Center along the Salmon Run Trail or park near the dam and pick up the trail at that location. The 5-mile Salmon Run Trail extends from the visitor center to the dam, passing through a number of the park's developed campgrounds as it winds among stands of ponderosa pine and Gambel oak. Because mountain bikers use this trail, be sure to step off the trail when you stop to look at a bird. While most of the trail has a level surface, there are sections that are uneven and hilly, with loose gravel, particularly between the Blanco and Brushy Point Campgrounds. It is rated as having moderate difficulty. A trail guide is available online at http://www.emnrd.state.nm.us/SPD/documents/SalmonRunTrailWebGuide.pdf.

The Rio Chama Trail starts at the parking area near the dam. As you begin this trail, you will walk down a wooden staircase and then across a suspension bridge over the Rio Chama. The trail extends 5 miles to El Vado State Park. If you visit before Memorial Day or after Labor Day, you will be able to walk only a short distance before coming to the Rio Chama Wildlife Management Area, which is closed to all except hunters during other times of the year.

As Heron is a no-wake lake, it is ideal for birding by canoe or kayak. This method of birding allows you to explore coves and see species that might not be visible from shore.

County: Rio Arriba

eBird Hotspots: Heron Lake SP

Web site: http://www.emnrd.state.nm.us/SPD/heronlakestatepark.html

Target Birds

Common Merganser This stately diving duck lives on the lake year-round and breeds at this location. In the spring you might see it flying toward its nesting cavity in a tree.

Common Loon While not present every year, it is a frequent visitor to the lake, both during the winter and in April and May.

Osprey Several pairs return to this location each year to breed. They arrive by early April and are present through August.

Heron Lake

Bald Eagle It is present all year, often hunting gulls during the summer. A pair nests nearby.

Western Grebe While only a few might be present during the winter, there have been several hundred during migration. If the water levels are suitable, it has been known to breed at the lake.

California Gull Normally present during most of the year, the first documented nesting occurred in 2013.

Cliff Swallow It nests along the cliffs just above the dam and is present from May through the end of July.

Other Birds

Species that can be seen at any time of year include Canada Goose, Mallard, Wild Turkey (East Meadow Trail), Great Blue Heron, Red-tailed Hawk, Ring-billed Gull, American Kestrel, Steller's Jay, Western Scrub-Jay, Black-billed Magpie, Clark's Nutcracker, American Crow, Common Raven, American Robin (primarily April through November), Mountain Chickadee, Juniper Titmouse, White-breasted and Pygmy Nuthatches, Western and Mountain Bluebirds, Red-winged Blackbird, Red Crossbill, and Evening Grosbeak.

During the summer, you can expect to see Ruddy Duck; Turkey Vulture; Killdeer; Spotted Sandpiper; Mourning Dove; Common Nighthawk; Black-chinned and Broad-tailed Hummingbirds; Northern Flicker; Western Wood-Pewee; Say's Phoebe; Ash-throated Flycatcher; Cassin's Kingbird; Plumbeous Vireo; Violet-green and Barn Swallows; Rock Wren; Yellow Warbler; Green-tailed and Spotted Towhees; Chipping, Vesper, and Lark Sparrows; Black-headed Grosbeak; Brewer's Blackbird; Brown-headed Cowbird; Bullock's Oriole; House Finch; Pine Siskin; and Lesser Goldfinch.

During migration, look for Gadwall; American Wigeon; Northern Shoveler; Green-winged Teal; Redhead; Common Goldeneye; Double-crested Cormorant (spring); American White Pelican; White-faced Ibis; Forster's Tern (spring); Red-naped Sapsucker (fall); Tree Swallow (spring); Blue-gray Gnatcatcher; Ruby-crowned Kinglet; Townsend's Solitaire; Yellow-rumped, Grace's, and Black-throated Gray Warblers; Townsend's and Wilson's Warblers (fall); Brewer's Sparrow; Western Tanager; Yellow-headed Blackbird; and Cassin's Finch.

Dark-eyed Junco is present primarily during the winter.

DIRECTIONS

From Santa Fe: From the intersection of I-25 (Exit 282) and St. Francis Drive (US 84/285), follow US 84 north approximately 97 miles (through the towns of Española and Tierra Amarilla toward Chama) to NM 95. Turn left and travel approximately 5.5 miles to the Heron Lake State Park Visitor Center.

From Chama: From the intersection of US 84/64 and NM 17, 1.5 miles south of town, travel south on US 84 approximately 10 miles to NM 95. Turn right and travel approximately 5.5 miles to the Visitor Center.

PARKING

Parking is available near the Visitor Center, marina, boat ramp, and both sides of the dam.

FEES

Day-use fee or state park annual pass

SPECIAL CONSIDERATIONS AND HAZARDS

See chapter 2 for further safety guidelines.

- Summer thunderstorms: Thunderstorms can come up suddenly during summer afternoons. Keep an eye on the sky, especially if you are birding by kayak.
- Winter driving conditions: US 84 and NM 95 can be snowpacked and icy during the winter. It is advisable to check the New Mexico Department of Transportation website (www.nmroads.com) or call the hotline (800–432–4269) for current road conditions during winter months.

FACILITIES

- Accessibility: All restrooms are accessible. The day-use areas are generally level. There are benches along the East Meadow and Salmon Run Trails. The Salmon Run Trail has some difficult sections, and the Rio Chama Trail is challenging.
- Restrooms: Located in campgrounds
- Water: Located in campgrounds
- Picnic tables: Near the Visitor Center
- Pets: Park rules require that dogs be on leash.
- Wi-Fi: This is one of the state parks that provides Wi-Fi service.

CAMPING

Several campgrounds are available in the park, ranging from primitive camping to full hookups. A number of the campgrounds are closed during the winter.

GAS, FOOD, AND LODGING

There is a restaurant just west of the park on NM 95. Gas and lodging are available at Stone House Lodge just west of the park on NM 95. Gas, food, and lodging are available in the Village of Chama, approximately 10 miles north of the intersection of NM 95 and US 84/64.

Cumbres and Toltec Railroad near Chama (photo by Christiana B. Burk)

Chama

Description

While the Village of Chama (7,892 feet), located in the foothills of the
Rocky Mountains, may be best known as the southern depot of the Cumbres and Toltec Scenic Railroad and well known by hunters and anglers, it is
also an area with a wide diversity of birds. It is one of the most reliable locations in north-central New Mexico for Lewis's Woodpecker.

Many of the target species can be seen by walking around the town. Most
of the Rio Chama is bordered by private property, so areas of access are limited. If you are staying at one of the motels south of town that border the
river, you can walk along the fishing trails to bird. The willow thickets attract a variety of species during spring through fall migration. There is also
public access to the river at an area managed by New Mexico Game and Fish
just south of the motels. It can be reached along a dirt road just north of the
RV park. This road runs between two fences: a well-maintained low fence
on the left and a second, less well-maintained 6-foot chain-link fence on the

right. The road continues for 200 or 300 yards and then turns sharply to the right toward the Chama. At this corner there is a New Mexico Department of Game and Fish sign identifying the spot as public fishing access. After approximately 100 yards, the road ends at a parking area under trees with room for two or three cars. You can walk the east side of the river going downstream about a half mile and upstream about a mile. The trail is dirt and stays close to the river. Look for posted signs to avoid trespass issues.

County: Rio Arriba

eBird Hotspots: Rio Chama Public Fishing Access

Target Birds

Great Blue Heron There is a rookery on the river near downtown where it nests.

Osprey It often can be seen flying over the river. A pair nests on a platform along the west side of US 84 just south of the intersection of CR 342 to Brazos.

Lewis's Woodpecker This is one of the most reliable locations for spotting it. Look for it in the cottonwoods along the village streets, as well as along the river south of town.

Black-billed Magpie It is very prevalent in town and along the river year-round.

Tree and Violet-green Swallows While both swallows nest in cavities near the river and can be seen swooping over the water between April and September, the Tree Swallow is the most prevalent.

Veery While not recorded every year, a few have nested along the Rio Chama, particularly south of town in June and July.

Gray Catbird It nests in the willow thickets along the river south of town. Look for it between May and September.

Evening Grosbeak While this roving species is possible at any time of year, it is most often seen in April, May, and August.

Other Birds

Species that can be seen at any time of year include Canada Goose, Mallard, Common Merganser, Red-tailed Hawk, Eurasian Collared-Dove, Belted Kingfisher, Downy and Hairy Woodpeckers, Northern Flicker, American Kestrel, Steller's Jay, Common Raven, American Crow, Black-capped Chickadee, White-breasted Nuthatch, American Dipper, Western and Mountain Bluebirds, American Robin, European Starling, Red-winged Blackbird, Western Meadowlark, and House Sparrow.

Lewis's Woodpecker

In the summer, look for Turkey Vulture, Cooper's and Swainson's Hawks, Spotted Sandpiper (along the river), Mourning Dove, Broad-tailed Hummingbird, Common Nighthawk, Western Wood-Pewee, Cordilleran Flycatcher, Say's Phoebe, Ash-throated Flycatcher, Plumbeous and Warbling Vireos, Cliff Swallow, House Wren, Cedar Waxwing, Yellow Warbler, Spotted Towhee, Song Sparrow, Black-headed Grosbeak, Brewer's Blackbird, Common Grackle, Brown-headed Cowbird, Bullock's Oriole, Pine Siskin, and American Goldfinch.

During migration you might spot White-faced Ibis; Rufous Hummingbird (late summer through fall); Red-naped Sapsucker; Ruby-crowned Kinglet; Northern Waterthrush; Yellow-rumped Warbler; Chipping, Lincoln's, and White-crowned Sparrows; Dark-eyed Junco; Western Tanager; and Cassin's Finch.

DIRECTIONS

From the intersection of I-25 (Exit 282) and St. Francis Drive (US 84/285) in Santa Fe, follow US 84 north approximately 108 miles (through the towns of Española and Tierra Amarilla) to where US 84 makes a sharp left. Instead, drive straight onto NM 17 for 1.6 miles to the Village of Chama.

PARKING

To bird around town, find a space along a side street. Do not park in the lot for the Cumbres and Toltec Scenic Railroad if you are not going to be a passenger. There is limited parking at the public fishing access south of town.

FEES

None

SPECIAL CONSIDERATIONS AND HAZARDS

See chapter 2 for further safety guidelines.

- Poison ivy: While not prevalent, it is possible along the river.
- Rattlesnakes: Keep your eyes and ears alert during warm weather.

FACILITIES

- Accessibility: Birding along the streets of the village is level. Fishing trails along the river can be uneven and may have obstacles, such as exposed roots.
- Restrooms: Public restrooms are 1.6 miles south of town in the Visitor Information Center (when open) at the junction of NM 17 and US 84 as it heads west.
- Water: Visitor Information Center or for sale in various businesses
- Picnic tables: Visitor Information Center
- Pets: While there are no restrictions on pets, it is always advisable to keep your pet on a leash.

CAMPING

There is an RV park south of town. The closest location for tent camping is at Heron Lake State Park.

GAS, FOOD, AND LODGING

There are ample options for gas, food, and lodging in the Village of Chama.

CHAPTER 8

Cochiti Lake Area

General Overview

This chapter describes four sites on Pueblo de Cochiti lands, as well as one in the rural Village of Peña Blanca situated between the Pueblo de Cochiti and the Santo Domingo (Kewa) Pueblo. Although a visitor can scan the landscape for birds and pull out on the road shoulder opposite the Osprey platform on NM 22, the portion of the road below the dam outlet along the Rio Grande is off-limits to the public. *Even though it is not posted, do not stop or park in the turnouts adjacent to the dam outlet or on the bridge itself.* Photography is allowed at all of the sites; however, no photography or the use of cell phones is allowed in the Pueblo de Cochiti.

Sites include areas along NM 22, both Cochiti Lake and Tetilla Peak Recreation Areas, and Kasha-Katuwe Tent Rocks National Monument. Birding habitat includes upland desert scrub, piñon-juniper, middle-elevation riparian, and agricultural.

Along NM 22 and Cochiti Highway
Description

NM 22 extends from I-25 (Exit 259) at Santo Domingo (Kewa) Pueblo and continues northwest to Cochiti Dam. At the dam, it turns sharply left to Pueblo de Cochiti, becoming Pueblo Route 85. Within the pueblo it intersects Pueblo Route 95, the road to Kasha-Katuwe Tent Rocks National Monument. Cochiti Highway continues from NM 22 at the dam, reaching Cochiti Golf Course, 3 miles north.

Leaving I-25, NM 22 crosses Santo Domingo Pueblo and enters the Village of Peña Blanca. Pueblo de Cochiti lands commence at the western edge of Peña Blanca. A number of locations along this route provide opportuni-

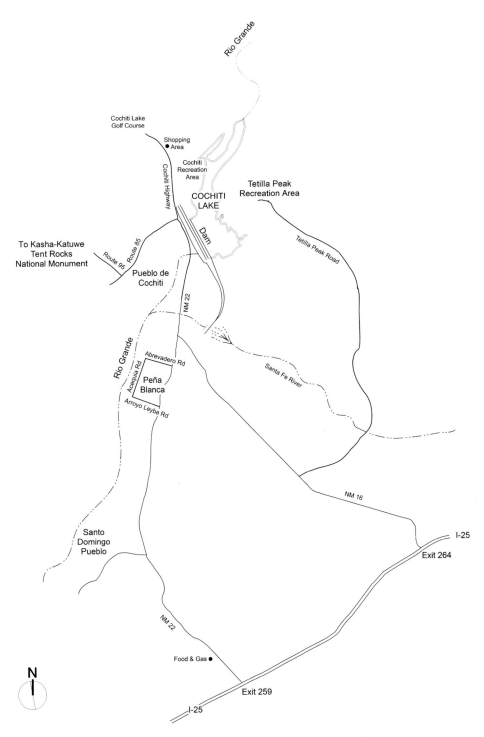

Rio Grande

Cochiti Lake
Golf Course

Shopping
Area

Cochiti
Recreation
Area

Tetilla Peak
Recreation Area

COCHITI
LAKE

Cochiti Highway

Tetilla Peak Road

To Kasha-Katuwe
Tent Rocks
National Monument

Route 95

Route 85

Dam

Pueblo de
Cochiti

NM 22

Rio Grande

Abrevadero Rd

Acequia Rd

Peña
Blanca

Santa Fe River

Arroyo Leyba Rd

NM 16

Santo
Domingo
Pueblo

I-25

Exit 264

NM 22

Food & Gas ●

Exit 259

I-25

N

Map 11. Cochiti Lake, Tetilla Peak, and Kasha-Katuwe Tent Rocks National Monument

ties to take advantage of specific bird species. In addition to the spots mentioned in this section, a separate site description is included for three locations: Cochiti Lake Recreation Area, Tetilla Peak Recreation Area, and Kasha-Katuwe Tent Rocks National Monument.

Start at I-25 (Exit 259) and travel northwest approximately 7.9 miles on NM 22 through Santo Domingo Pueblo to the Village of Peña Blanca.

Peña Blanca: Turn left on Arroyo Leyba Road and drive west about 0.8 mile to Acequia Road. Scan the fields and trees beyond the arroyo on the left and right without invading the privacy of the farm and homeowners. Parking is available near the Middle Rio Grande Conservancy District's (MRGCD) gate that provides access to the *bosque* (Spanish for "woodlands"). Access beyond the gate requires permission (call 505–247–0234). Continue 0.6 mile north along Acequia Road to Abrevadero Road to return to NM 22. During winter, birding along this drive can produce Red-tailed Hawk, Northern Harrier, Western and Mountain Bluebirds, and wintering sparrows. In spring, Ring-billed and Franklin's Gulls are often attracted to flooded fields. Migrating birds, such as Chipping and Savannah Sparrows, as well as Yellow-headed Blackbirds, are often seen in the area habitats in the fall. American Kestrel, Black-billed Magpie, Red-winged Blackbird, and Western Meadowlark are present year-round. *Please do not block private driveways when stopping to look at a bird. Pull off onto the shoulder to allow safe passage for local vehicles.*

As you continue north 4.3 miles, the highway intersects NM 16 (route to Cochiti Lake's Tetilla Recreation Area). Shortly after this junction the road crosses over a small wetlands area where the Santa Fe River joins the Rio Grande a short distance to the west. The bridge does not have a shoulder, and there is no safe place to park for birding.

Osprey platform: Look for an Osprey platform on a cluster of power poles 1.4 miles north of the intersection of NM 22 and NM 16. The Osprey return each spring in early April and often can be seen on the platform before actual nesting begins and then throughout the nesting season. The shoulder is wide enough opposite the platform to pull off and stop.

The right turn to the access road to Cochiti Lake Recreation Area is a short distance beyond the left turn to the Pueblo and Kasha-Katuwe Tent Rocks National Monument.

Town of Cochiti Lake shopping area: A small shopping complex on the right 0.4 mile past the entrance to Cochiti Lake Recreation Area has a gas

Osprey platform along NM 22

station and convenience store. The wooded area along the arroyo is usually productive for birds.

Cochiti Golf Course Clubhouse: The golf course is 1.8 miles beyond the shopping area. A Canyon Towhee will undoubtedly greet you in the parking lot. Grab a bite to eat at the grill and take it out to the picnic tables to watch birds coming to drink at the water feature and at the edge of the golf course. Western Bluebirds, and sometimes Evening Grosbeaks, can be seen.

County: Sandoval

eBird Hotspots: No

Target Birds

Pinyon Jay While unpredictable, flocks of these raucous jays are often found in piñon-juniper habitat near the Town of Cochiti Lake.

Black-billed Magpie
It is fairly reliable on the farms in Peña Blanca that back up to the bosque.

Osprey Osprey can be seen on or near the nesting platform between early April and early October.

Lewis's Woodpecker While not regular, it is often seen in Peña Blanca.

American Kestrel It is found year-round in Peña Blanca.

Western and Mountain Bluebirds These species migrate from higher elevations in late September and can be seen at any of the sites mentioned in this section through late March or early April.

Townsend's Solitaire It also migrates from higher elevations in mid- to late September and winters in piñon-juniper habitat near the Town of Cochiti Lake.

Brewer's Blackbird Look for flocks on the farms of Peña Blanca from the end of September through early May, especially those with animals.

Evening Grosbeak While unpredictable, it is often seen near the Town of Cochiti Lake or Cochiti Golf Course.

Other Birds

During late fall, winter, and early spring, look for Northern Harrier; Sharp-shinned Hawk; Bald Eagle; Sandhill Crane; Ring-billed Gull; Ruby-crowned Kinglet; Yellow-rumped Warbler; Savannah, Song, Lincoln's, and White-crowned Sparrows; and American Goldfinch.

During spring and/or fall migration you might see Killdeer; Franklin's Gull; Violet-green Swallow; and Brewer's, Chipping, and Lark Sparrows.

Summer residents include Swainson's Hawk, Common Nighthawk, Cattle Egret, Green Heron, Western Kingbird, Barn Swallow, Northern Mockingbird, Black-headed and Blue Grosbeaks, Brown-headed Cowbird, and Bullock's Oriole.

Year-round birds include Cooper's and Red-tailed Hawks, Rock Pigeon, Eurasian Collared-Dove, White-winged and Mourning Doves, Belted Kingfisher, Ladder-backed and Downy Woodpeckers, Northern Flicker, Black and Say's Phoebes, Loggerhead Shrike, Western Scrub-Jay, American Crow, Common Raven, Black-capped Chickadee, Bushtit, White-breasted Nuthatch, Bewick's Wren, American Robin, European Starling, Spotted Towhee, Red-winged Blackbird, Western Meadowlark, Great-tailed Grackle, House Finch, Pine Siskin, Lesser Goldfinch, and House Sparrow.

DIRECTIONS

Birding along NM 22 begins from I-25 (Exit 259/Santo Domingo [Kewa] Pueblo). The exit is approximately 23 miles southwest of the intersection of I-25 (Exit 282) and US 285/84 in Santa Fe.

Exit 259 is 33 miles north of the intersection of I-25 and I-40 in Albuquerque.

PARKING

The only parking in Peña Blanca is at the west end of Arroyo Leyba Road next to the MRGCD gate. Please do not park along the back roads or block driveways. Parking is available on the shoulder opposite the Osprey platform. There is parking at the Cochiti Lake shopping center and at the Cochiti Golf Course.

FEES

None

SPECIAL CONSIDERATIONS AND HAZARDS

See chapter 2 for further safety guidelines.

- Traffic: Watch for fast-moving vehicles if you get out of your car at the Osprey platform along NM 22.
- Private property: Please do not intrude on the privacy or property of the residents of Peña Blanca.
- Pueblo de Cochiti: Some areas are posted as off-limits and restrict photography and the use of cell phones.

FACILITIES

- Accessibility: Most birding along this route is by car. The arroyo adjacent to the shopping center in the Town of Cochiti Lake is slightly hilly, and there is no trail. The Cochiti Golf Course Clubhouse area is level.
- Restrooms: Restrooms are available at the convenience stores on NM 22 at the I-25/Santo Domingo (Kewa) Exit (259), the Cochiti Lake Recreation Area, and the Cochiti Golf Course Clubhouse.
- Water: Water is available at Cochiti Lake Recreation Area campground or can be purchased at the general store in Peña Blanca, at a mini-mart in the Town of Cochiti Lake, or at the grill at the Cochiti Golf Course.
- Picnic tables: Available at Cochiti Lake Recreation Area and at Cochiti Golf Course when food is purchased

CAMPING

There is a campground at Cochiti Lake Recreation Area.

GAS, FOOD, AND LODGING
There are a gas station and convenience store at the I-25/Santo Domingo (Kewa) Exit (259) and at the Town of Cochiti Lake. Food can be purchased at the Peña Blanca general store. There is a restaurant at Cochiti Golf Course. Nearest lodging is in Santa Fe.

Cochiti Lake Recreation Area
Description
Located on land belonging to the Pueblo de Cochiti, the lake and surrounding recreation area are managed by the US Army Corps of Engineers. The dam on the Rio Grande that creates Cochiti Lake (reservoir) is one of the 10 largest earthfill dams in the United States. The depth of the lake, as well as its "no-wake" designation, attracts a variety of waterfowl year-round, particularly during the winter months. Even when the lake is frozen, there may be good opportunities to view Bald Eagle perched on the ice.

Begin by stopping at the Visitor Center (during winter months it is open only on weekends). In addition to exploring the informational displays inside, take time to scan the shrubs and trees in the arroyo below and around the parking area. If the Visitor Center is not open, the overlook between the Visitor Center and the boat ramp provides a similar vantage point.

There are two access areas for the lake: near the boat ramp and from the swim beach (accessible year-round on foot; closed to vehicles October 15–April 15). Start by driving down Cochiti Lake Road toward the boat ramp to one of the parking areas on either side of the ramp. Waterfowl can best be observed with a scope, as these species are often clustered on the far side of the lake. Check on the hillside across the cove from the picnic area for Rock Wren. As you drive back up the road, look for the turnoff to the swim beach. The beach is closer to the dam itself, and waterfowl often congregate in that area. The swim beach is also a gathering area for gulls and occasionally shorebirds when human activity is not disturbing them.

After returning to the boat ramp road from the swim beach, follow the signs to the picnic area. Park and explore the piñon-juniper habitat.

The reservoir can also be explored from the water by canoe, kayak, or no-wake boat launched from the ramp. For the most positive aquatic birding experience, an early-morning start is recommended, navigating counterclockwise and making sure not to drift too near the birds.

County: Sandoval

Birders at Cochiti Recreation Area

eBird Hotspots: Cochiti Lake
website: http://www.spa.usace.army.mil/Missions/CivilWorks/Recreation/
CochitiLake.aspx
Facebook: https://www.facebook.com/USACE.Cochiti.Lake

Target Birds
Bufflehead Sometimes only a few are present, but often between mid-October
and mid-April rafts of Bufflehead can be seen on the lake.
Common Merganser It is present from early October through the first week in
May, with the greatest prevalence between November and February.
Eared Grebe Although one or more might be seen from September through
March, the greatest numbers are encountered in April.
Western and Clark's Grebes While both species of grebe are present year-round,
Western Grebe is the most prevalent.
Osprey A pair of Ospreys nest in the area and usually can be seen from the
last week in March through the third week in October.
Bald Eagle Several winter in the area from December through February.

Ring-billed Gull This gull winters on the lake, arriving at the end of September, and is present through early May.

Rock Wren It is a year-round resident (rare in winter) and can be seen and heard along the rocky cliffs on the west side of the lake.

Western and Mountain Bluebirds Both species begin to migrate from higher elevations to the piñon-juniper habitat in the recreation area in September and are present through late April.

Townsend's Solitaire It also migrates from higher elevations to piñon-juniper habitat in late September and remains through mid-April.

Other Birds

During late fall, winter, and early spring, Canada and sometimes Cackling Geese can be found on or near the lake, along with Gadwall, American Wigeon, Northern Shoveler, Northern Pintail, Green-winged Teal, and Lesser Scaup. Not as prevalent, but present at some times during the winter, are Canvasback, Redhead, Ring-necked and Ruddy Ducks, Common Goldeneye, and Hooded and Red-breasted Mergansers. Other winter visitors include Northern Harrier, Sharp-shinned and Red-tailed Hawks, Peregrine Falcon, Northern Flicker, Ruby-crowned Kinglet, American Robin, Song and White-crowned Sparrows, Dark-eyed Junco, and Pine Siskin.

During spring and/or fall migration, look for Cinnamon Teal, Double-crested Cormorant, Black-crowned Night-Heron, Turkey Vulture, Sandhill Crane, Franklin's Gull, White-throated Swift, Tree and Bank Swallows, American Pipit, Yellow-rumped and Wilson's Warblers, and Chipping and Brewer's Sparrows.

Summer residents include Cassin's and Western Kingbirds; Northern Rough-winged, Violet-green, Barn, and Cliff Swallows; Yellow Warbler; Yellow-breasted Chat; Lark Sparrow; Black-headed and Blue Grosbeaks; and Brown-headed Cowbird.

Mallard, Great Blue Heron, American Coot, American Kestrel, Black and Say's Phoebes, Loggerhead Shrike, Western Scrub-Jay, American Crow, Common Raven, Bushtit, Canyon and Bewick's Wrens, Red-winged Blackbird, and Lesser Goldfinch are present year-round.

DIRECTIONS

From Santa Fe: At the intersection of I-25 (Exit 282) and US 285/84 (St. Francis Drive), travel south approximately 17 miles to Cochiti Pueblo (Exit

264). Turn right on NM 16 and travel northwest 8.2 miles to NM 22. Turn right on NM 22, traveling past the dam outlet. After 2.7 miles, NM 22 makes a sharp left toward Pueblo de Cochiti. Instead, continue straight 1 mile. The entrance to Cochiti Lake Recreation Area is on the right.

From Albuquerque: The I-25 Cochiti Pueblo Exit (264) is approximately 38.5 miles north of the intersection of I-25 and I-40 in Albuquerque.

PARKING

Parking is available at the Visitor Center, overlook, boat ramp, swim beach, and Cochiti Picnic Area.

FEES

There are fees for camping and use of boat-launching facilities. No fees are charged for canoeing or kayaking.

SPECIAL CONSIDERATIONS AND HAZARDS

See chapter 2 for further safety guidelines.

■ Restricted Areas: The recreation area is situated on Pueblo de Cochiti lands. Some areas are posted as off-limits.

■ Photography: No photography is allowed below the dam; however, photography is allowed in the recreation area.

■ Alcohol: No alcohol is allowed anywhere on Pueblo de Cochiti lands, including the recreation area.

■ Winter weather: Check road and lake conditions during the winter. The surface of the lake is often frozen from mid-November through mid-February.

FACILITIES

■ Accessibility: The majority of the paths within the recreation area are level.

■ Restrooms: Visitor Center, overlook, boat ramp, swim beach, and Cochiti Picnic Area

■ Water: Drinking fountain inside Visitor Center and from faucet in campground

■ Picnic tables: Overlook, boat ramp, swim beach, and Cochiti Picnic Area

■ Pets: Recreation Area rules require dogs to be on leash.

There is a campground on-site.

There are a gas station and convenience store at the Town of Cochiti Lake and a restaurant at Cochiti Golf Course. Nearest lodging is in Santa Fe.

Tetilla Peak Recreation Area
Description

Situated on Pueblo de Cochiti lands, Tetilla Peak Recreation Area, managed by the US Army Corps of Engineers, is a seasonal facility (April 15–October 15) located on the northeast side of Cochiti Lake. Begin birding after exiting I-25 (Exit 264) on NM 16. Grassland birds can be seen flying or on fence lines in this plains-grassland and upland desert scrub habitat. Shortly after turning right (north) onto Tetilla Peak Road, you will reach the Santa Fe River. There is a wide shoulder on the right after crossing the bridge, where you can park to scan the riparian habitat. The seasonal Tetilla Peak Road follows a volcanic escarpment, allowing the habitat to gradually change from desert scrub to piñon-juniper as the road gains elevation. There are patches of cholla cactus along the base of the cliffs where Black-throated Sparrows may be found. In many locations, the shoulder is wide enough to pull off to allow birding. Because the road crosses the Pueblo de Cochiti, it is important to stay on the road shoulder while birding.

After entering the Recreation Area, turn right toward the overlook. Check for birds along the periphery of the parking area and over at the campground. The Russian olive and other trees around the campground restroom are often productive. From the overlook viewing platform, scan Cochiti Lake and upstream on the Rio Grande. A rocky trail starting near the observation platform leads north along the top of the river gorge, providing a view of the sandbars. You can also walk north from the parking lot on a service road to use a spotting scope to survey the Rio Grande.

Next, drive down the road toward the boat launch and turn left just before the ramp on a graded dirt road. The sign will indicate that it leads to the windsurfing area. Follow the graded road to its end near the east end of the lake. There is a large area where you can park and walk to the edge of the lake. This spot can be busy with human activity after 9:30 a.m., so your best chance of seeing migrating shorebirds when present is to arrive as early

Cochiti Lake from Tetilla Overlook

as possible. As you return, make a left turn at the vault toilet and drive along a short unpaved road between a deciduous grove and the lake. You can park alongside the woodland to look for birds in the cottonwoods, as well as in the vegetation along the lake. This area is particularly productive in September during fall migration. Yellow-breasted Chats and Blue Grosbeaks are numerous in summer. Also stop at the paved parking lot with picnic tables and restroom. From the floating dock, there is a good view of the lake and nearby cove. As you head back, explore the dirt roads that lead off to the left. The area can also be explored from the water by canoe, kayak, or no-wake boat. You can launch from the ramp at the Cochiti Lake Recreation Area (when the Tetilla Peak Recreation Area is closed) or the ramp at Tetilla Peak Recreation Area in order to explore the coves along this side of the lake and into the mouth of the Rio Grande.

County: Sandoval

eBird Hotspots: Tetilla Recreation Area

website: http://www.spa.usace.army.mil/Missions/CivilWorks/Recreation/CochitiLake.aspx

Facebook: https://www.facebook.com/USACE.Cochiti.Lake

Target Birds

Scaled Quail It can be found on the hillsides on either side of the Rio Grande near the overlook.

Eared Grebe Eared Grebe is prevalent on the reservoir during spring and fall migration and is more easily seen from the Tetilla Peak Recreation Area. Good locations for spotting one are at the picnic area with the floating dock and the windsurfing launch area.

Western and Clark's Grebes Although Western Grebe is dominant, both of these grebe species are present year-round and can be seen more easily from the Tetilla Peak Recreation Area side of the lake.

Osprey Osprey nests in the area and may be observed fishing over the lake from early April through the end of September or early October.

Cassin's and Western Kingbirds Arriving in late April, the Cassin's Kingbird nests in the cottonwood woodlands and piñon-juniper habitat, staying through the end of September. Western Kingbird nests in the cottonwoods as well.

Sage Thrasher It migrates through this area in September and may be spotted along the Tetilla Peak Road and within the Recreation Area, particularly around the overlook.

Loggerhead Shrike It frequently is perched along the Tetilla Peak Road.

Other Birds

During spring, a variety of waterfowl frequent the lake and upper Rio Grande, including Gadwall; Mallard; Blue-winged, Cinnamon, and Green-winged Teal; Northern Shoveler; Lesser Scaup; Bufflehead; and Hooded and Common Mergansers. The prevalence of any of these species will depend on migration patterns in any given year. Raptors include Turkey Vulture, Cooper's and Swainson's Hawks that breed in the area, as well as migrating Red-tailed Hawk. Migrating shorebirds include Killdeer, American Avocet, Spotted and Least Sandpipers, Willet, Marbled Godwit, and Long-billed Dowitcher. Gulls include Franklin's, an occasional California, as well as lingering Ring-billed. Migrating songbirds include Northern Rough-winged, Tree, Violet-green, Bank, and Cliff Swallows; Orange-crowned, Virginia's, and Yellow-rumped Warblers; and Chipping, Brewer's, Lark, and Savannah Sparrows.

During the summer you can see Turkey Vulture and Swainson's Hawk soaring. Near or in the cottonwoods along the lake, expect to find Spotted Sandpiper, Black-chinned Hummingbird, Barn and Cliff Swallows, Western Wood-Pewee, Ash-throated Flycatcher, Yellow Warbler, Yellow-breasted

Chat, Lark Sparrow, Blue Grosbeak, Indigo Bunting, and Bullock's Oriole. Along Tetilla Peak Road, look for Northern Mockingbird and Black-throated Sparrow.

Fall and early-winter species include Gadwall; Mallard; American Wigeon; Northern Pintail; Bufflehead; Common Merganser; American Coot; Williamson's and Red-naped Sapsuckers; White-throated Swift; Violet-green Swallow; Horned Lark; Western Bluebird; American Robin; Orange-crowned, Yellow-rumped, and Wilson's Warblers; Chipping, Brewer's, Vesper, Song, and White-crowned Sparrows; Dark-eyed Junco; Western Meadowlark; Brewer's Blackbird; and Pine Siskin.

Year-round residents that can be observed when the area is open include Mourning Dove, Ladder-backed Woodpecker, Northern Flicker, Say's Phoebe, Loggerhead Shrike, Pinyon Jay, Western Scrub-Jay, Common Raven, Curve-billed Thrasher, Canyon Towhee, and House Finch.

DIRECTIONS

From Exit 264 on I-25 (approximately 30 miles southwest of Santa Fe or 40 miles north of Albuquerque), take NM 16 northwest approximately 4 miles to Indian Service Route 841. Turn right. Shortly after crossing the Santa Fe River, the road becomes Tetilla Peak Road (gate will be locked when Recreation Area is closed October 15–April 15). The road ends at the Recreation Area.

PARKING

Although cars can be parked at a number of locations, there are only two formal parking areas: Overlook and Picnic Area.

FEES

Fees are charged for camping and use of boat-launching facilities. Kayak and canoe launching is free.

SPECIAL CONSIDERATIONS AND HAZARDS

See chapter 2 for further safety guidelines.

- Rattlesnakes: Rattlesnakes may be near the overlook area.
- Weekend activities: Heavy boating/fishing/windsurfing traffic occurs on weekends and holidays by 9:30 a.m.
- Restricted areas: The Recreation Area is situated on lands of the Pueblo de Cochiti. Some areas are posted as off-limits.

- Alcohol: No alcohol is allowed anywhere on lands of the Pueblo de Cochiti, including the Recreation Area.

- Accessibility: Most birding areas are fairly level, with the exception of the trail that leads down to the lake from the overlook and upstream along the Rio Grande.
- Restrooms: Several restrooms are available throughout the area.
- Water: Campground and overlook
- Picnic tables: There are picnic tables at an area along the road that leads east from the boat ramp.
- Pets: Recreation Area rules require that dogs be on leash.

CAMPING

There is a campground on-site.

GAS, FOOD, AND LODGING

There are a gas station and convenience store at the Town of Cochiti Lake and a restaurant at Cochiti Golf Course. These are located on the other side of the lake. See the section on Cochiti Lake Recreation Area for directions. Nearest lodging is in Santa Fe.

Kasha-Katuwe Tent Rocks National Monument

Description

Kasha-Katuwe Tent Rocks National Monument is an extraordinary geologic location that also provides an opportunity to encounter piñon-juniper bird species. The rock formations at the site are composed of volcanic debris layers 400 feet thick. They were deposited 6 or 7 million years ago by a series of pyroclastic flows and surges of superheated volcanic ash and then eroded over time into distinctive tent-shaped hoodoos and pedestals. Hoodoos are erosional cones or columns formed when an erosion-resistant caprock exists on top of softer deposits, in this case, volcanic tuff. Some of the hoodoos at Kasha-Katuwe ("white cliffs" in Keresan, the traditional language of the Pueblo de Cochiti) are as tall as 90 feet above the trail. Administered by the Bureau of Land Management in cooperation with the Pueblo de Cochiti, Kasha-Katuwe is designated an Area of Critical Environmental Concern.

Kasha-Katuwe Tent Rocks National Monument

Although the piñon-juniper habitat near the parking area and the Cave Loop Trail may be better birding areas, the Canyon Trail, a 1.1-mile (one way) spur trail leading into a narrow, winding slot canyon, affords the most spectacular views and allows up-close inspection of an otherworldly landscape. In the slot canyon it is possible to see Canyon Wren or White-throated Swift, Violet-green Swallow, and Red-tailed Hawk soaring overhead. The trail leading from the parking lot and the Cave Loop Trail can produce Western Scrub-Jay, Juniper Titmouse, Canyon Towhee, and other species expected in piñon-juniper habitat.

With a high-clearance vehicle, you can continue past the main parking area and up BLM Route 1011 (unpaved) to the Veterans Memorial Scenic Overlook. Otherwise, you can walk along the road for a short distance for excellent views of the rock formations. The gravel road leads through a canyon with piñon-juniper habitat and has more vegetation than the main part of the monument. The views from the overlook are similar to those from the top of the Canyon Trail. There is a 1-mile level gravel trail that loops along the top of the plateau at the Veterans Memorial.

Winter hours (November 1–March 10): 8:00 a.m. to 5:00 p.m.; gates locked at 4:00 p.m. Summer hours (March 11–October 31): 7:00 a.m. to 7:00 p.m.; gates locked at 6:00 p.m. *Visitors must be out by closing time.* The monument is closed on Thanksgiving, Christmas Eve, Christmas Day, and New Year's Day.

County: Sandoval

eBird Hotspots: Kasha-Katuwe Tent Rocks NM

website: http://www.blm.gov/nm/st/en/prog/recreation/rio_puerco/ kasha_katuwe_tent_rocks.html

Target Birds

White-throated Swift Look for swifts soaring above the canyon tops from late March to early October.

Gray Flycatcher It nests in the piñon-juniper habitat along the road to the Veterans Memorial Overlook and on the plateau.

Juniper Titmouse Listen for its soft tapping and calls in juniper trees along the trails year-round.

Rock Wren Rock Wren can be seen bobbing among the rocks along the Cave Loop Trail year-round.

Canyon Wren Listen for the descending call of this wren in the slot canyon year-round.

Hepatic Tanager This species can be found in piñon-juniper habitat near the cliffs from May through September.

Other Birds

During winter months, look for Red-tailed Hawk, Mountain and Western Bluebirds, Townsend's Solitaire, and Dark-eyed Junco.

Species that spend the summer at this location include Turkey Vulture, Say's Phoebe, Ash-throated Flycatcher, Cassin's Kingbird, Violet-green Swallow, Chipping Sparrow, and Lesser Goldfinch.

Year-round species include Hairy Woodpecker, Northern Flicker, Western Scrub-Jay, Common Raven, Bushtit, Bewick's Wren, Spotted and Canyon Towhees, and House Finch.

DIRECTIONS

From Santa Fe: At the intersection of I-25 (Exit 282) and US 84/285 (St. Francis Drive), travel southwest approximately 17 miles to Cochiti Pueblo (Exit 264). Turn right on NM 16 and travel northwest 8.2 miles to NM 22.

Turn right on NM 22, traveling past the dam outlet. After 2.7 miles, NM 22 makes a sharp left toward Pueblo de Cochiti. Turn left and continue on NM 22, which shortly becomes Pueblo Route 85. The sign to Kasha-Katuwe is on the right.

From Albuquerque: The I-25/NM 16 Cochiti Pueblo Exit (264) is approximately 38.5 miles north of the intersection of I-25 and I-40 in Albuquerque.

PARKING

There is a fairly large parking lot at the trailhead in the main part of the monument. It can be very crowded on weekends. There is also a large parking area at the Veterans Memorial Overlook.

FEES

$5.00 per vehicle or federal pass

SPECIAL CONSIDERATIONS AND HAZARDS

See chapter 2 for further safety guidelines.

- Special requirements: Please respect the traditions and privacy of the Pueblo de Cochiti.
- Photography: Photography, drawings, and recordings are not permitted in the Pueblo or on tribal land; however, photography is allowed at the national monument. Permits are required from BLM for commercial filming.
- Site of geologic and cultural significance: Please stay on the designated trails. Collecting plants, rocks, or obsidian "apache tears" is prohibited.
- Unannounced closings: The national monument is located on Pueblo de Cochiti land. The site may be closed by order of the Cochiti tribal governor, in which case a notice will be posted at the gate.
- Road conditions: During periods of inclement weather, the access road may wash out or become impassible.
- Winter: The Canyon Trail may become an icy "sled track."
- Summer weather: There is the potential for thunderstorms, which can lead to flash flooding in the slot canyon or in arroyos that cross the access road.

- Accessibility: The right-hand section of the Cave Loop Trail is level and accessible to all (0.5 mile). When the trail turns west, it becomes more rugged and has sections that are steep with loose gravel on the descent. The Canyon Trail is fairly flat, although there are some places where you have to scramble up rocks along the first portion. However, the final stretch of the trail emerges from the slot canyon up a steep (630-foot) slope with loose gravel leading to the mesa top. There is a paved path to the overlook viewing area at the Veterans Memorial Overlook, as well as a level gravel trail.
- Restrooms: Available at both locations
- Water: None available
- Picnic tables: Available at both locations
- Pets: Dogs are *not* allowed on any trails.

CAMPING

This is a day-use area. Camping is available at Cochiti Lake Recreation Area.

GAS, FOOD, AND LODGING

There are a gas station and convenience store at the Town of Cochiti Lake and a restaurant at Cochiti Golf Course. Nearest lodging is in Santa Fe.

CHAPTER 9

Along the Upper Pecos River

General Overview

The section of the Pecos River from its headwaters arising high in the Pecos Wilderness in the Sangre de Cristo Mountains to the Terrero area has been designated a Wild and Scenic River by Congress for the purpose of protecting naturally flowing rivers from development that would substantially change their wild or scenic nature. The river flows south through a narrow, rugged gorge that gives way to a broader canyon. When it arrives at the Village of Pecos, it meanders through a broad valley and crosses I-25, where it once again enters a steep-walled canyon. Although the Pecos River makes its way across eastern New Mexico and flows into Texas, the sites described in this chapter follow the river only as far south as Villanueva State Park.

The six sites include one collective description that highlights several locations along the river north of the Village of Pecos that are productive for birding. In addition, we have included a site near Terrero, one at the upper reaches of the canyon at 8,900 feet, a site at the bottom of the canyon just north of the town of Pecos, Pecos National Historical Park, and Villanueva State Park. These sites provide a variety of habitats in addition to the montane riparian habitat along the river, including mixed conifer forest, ponderosa pine, and piñon-juniper.

Although NM 63 is open during the winter, the sites north of the Village of Pecos are not open during the winter. Pecos National Historical Park is open year-round, but winter weather conditions may affect its hours. Villanueva State Park has seasonal closures. See individual site descriptions for more complete information on winter closings.

Pecos National Historical Park

Description

Located on the historic Santa Fe Trail in the Pecos Valley at 6,930 feet, the Pecos National Historical Park has a broad array of piñon-juniper species, in addition to historical and cultural interpretation of the Pecos Valley.

After checking in at the Visitor Center to pay your fee and pick up a brochure, you can walk the 1.25-mile (round-trip) paved Ruins Trail. Check for birds in the ruins and in the grassy fields beyond the low stone wall to your right. At the overlook, scan the riparian habitat along Glorieta Creek below. As you return to the Visitor Center, the trail goes over a boardwalk that straddles wagon ruts of the Santa Fe Trail.

The park is open daily, except Thanksgiving, Christmas, and New Year's Day, from 8:00 a.m. to 6:00 p.m. during the summer (Memorial Day through Labor Day) and 8:00 a.m. to 4:30 p.m. in the winter.

County: San Miguel

eBird Hotspots: Pecos NHP

Target Birds

Say's Phoebe This dry-habitat phoebe often can be seen hovering over the fields while searching for insects during spring and summer.

Ash-throated Flycatcher Arriving in early May, it can be seen through mid-August. Look for the pale throat and rufous-colored tail.

Cassin's Kingbird It arrives about the same time as the Ash-throated Flycatcher and can be found in areas of the park where there are trees.

Pinyon Jay While it might be seen in large flocks at any time of year, it is most prevalent during spring, fall, and winter.

Rock Wren Watch for it bobbing on rocky outcroppings or near the ruins from spring through early fall.

Townsend's Solitaire It arrives at the end of August to set up its winter feeding territory and is present until the end of March/early April.

Other Birds

Species that can be seen at any time of year include Red-tailed Hawk, Ladder-backed Woodpecker, Northern Flicker, American Kestrel, Western Scrub-Jay, Black-billed Magpie (infrequent), American Crow, Common Raven, Juniper Titmouse, Bushtit, Bewick's Wren, Mountain and Western Bluebirds, Western Meadowlark, House Finch, and House Sparrow.

Pecos National Historical Park Trail

During the summer months, look for Turkey Vulture, Mourning Dove, Common Nighthawk (dusk and dawn near entrance road), Broad-billed Hummingbird, Western Kingbird, Barn and Cliff Swallows, Northern Mockingbird, Spotted and Canyon Towhees, Lark Sparrow, Black-headed and Blue Grosbeaks, and Lesser Goldfinch.

Possibilities during migration include Red-naped Sapsucker; Western Wood-Pewee; Northern Rough-winged and Violet-green Swallows; Blue-gray Gnatcatcher; Ruby-crowned Kinglet; Orange-crowned, Yellow-rumped, and Wilson's Warblers; Chipping and Vesper Sparrows; Brewer's and Clay-colored Sparrows (occasional); and Pine Siskin.

Wintering species include White-crowned Sparrow and Dark-eyed Junco (Gray-headed and Pink-sided subspecies).

DIRECTIONS

From Santa Fe: From the intersection of I-25 (Exit 282) and St. Francis Drive (US 84/285), take I-25 northbound (heading southeast) 16.5 miles to Exit 299 (Glorieta). Cross the overpass and take NM 50 east approximately 6 miles to the Village of Pecos. Turn south onto NM 63 and travel about 2 miles to the park entrance.

From Las Vegas, New Mexico: From I-25 Exit 345 (NM 104), travel south on I-25 approximately 38 miles to Exit 307 (Rowe). Take NM 63 north 3.5 miles to the park entrance.

PARKING

There is a large lot near the Visitor Center.

FEES

$3.00 (good for one week) or federal pass

SPECIAL CONSIDERATIONS AND HAZARDS

See chapter 2 for further safety guidelines.

- Rattlesnakes: Rattlesnakes are possible. Keep your eyes and ears open as you walk along the trail.
- Archaeological sensitivity: This historical park contains artifacts from past inhabitants. Because it is a national park site, nothing should be disturbed in any way. Stay on the trails. The area is protected under the Archaeological Resources Protection Act.

FACILITIES

- Accessibility: The trail leading around the ruins is mostly paved with an easy grade.
- Restrooms: Accessible restrooms are adjacent to the Visitor Center and next to the headquarters building.
- Water: Drinking fountains are located at the Visitor Center and next to the headquarters building.
- Picnic tables: A latilla-covered picnic area is adjacent to the Visitor Center, and there is a picnic area adjacent to the upper parking lot.
- Pets: Leashed pets are allowed on trails. Pets are not permitted inside public buildings with the exception of service animals on a leash.

CAMPING

There are a variety of campgrounds along NM 63 in the national forest.

GAS, FOOD, AND LODGING

Gas and food are available in the Village of Pecos. Bed-and-breakfasts, guest ranches, and cabins are available at various locations in the general area of

Pecos. There are motels and hotels in Las Vegas and Santa Fe, both about 30 minutes away, in opposite directions on I-25.

Description

Monastery Lake (elevation 7,026 feet), located just north of the Village of Pecos, is leased by New Mexico Department of Game and Fish (NMDGF) from the Pecos Benedictine monastery for fishing. Regulations on its use may be subject to change. Please see NMDGF regulations for what is currently allowed on this property (http://www.wildlife.state.nm.us/recreation/hunting/open_gate/MonasteryLake.htm). The area is open from sunrise to sunset.

The lake and adjacent Pecos River also provide good habitat for birding. Because this area is extremely popular with anglers, the best time to bird is very early in the morning and during the middle of the week.

Begin your birding as you turn left off NM 63 and head toward the parking lot. The field on the left can be particularly productive in late summer/early fall for migrating sparrows. Also check the trees on your right.

After parking, follow the trail that leads from the parking lot straight ahead (west) toward the river to survey species in that area. Return partway and turn right on the first trail that leads along the west side of the lake. You can either walk around the lake or return along the same route. It is recommended to take one of the trails through the riparian area adjacent to the parking lot.

County: San Miguel

eBird Hotspots: Monastery Lake

Target Birds

Osprey A pair nests in the area and often can be seen fishing in the lake between April and September.

Belted Kingfisher It can be seen fishing at both the lake and along the river.

Western Wood-Pewee It arrives in May and is present through mid-September.

Black-billed Magpie A year-round resident, it has been known to breed in the area.

Townsend's Solitaire It descends from higher elevations in late August and spends the winter at this location. It can be heard singing when it arrives in late summer as it establishes its winter feeding territories.

Gray Catbird Look for it lurking in the dense understory next to the river between June and October. It breeds at this location.

Monastery Lake

Song Sparrow This also is a species of the dense understory next to the river and lake. It is a common breeder here, and some can be found here year-round.

Other Birds
Species that can be seen at any time of year include Mallard, Great Blue Heron, Pied-billed Grebe, Cooper's and Red-tailed Hawks, Hairy Woodpecker, American Crow, Common Raven, Black-capped and Mountain Chickadees, White-breasted and Pygmy Nuthatches, Western Bluebird, American Robin, Red-winged Blackbird, House Finch, and Evening Grosbeak.

Summering birds include Mourning Dove; Black-chinned and Broad-tailed Hummingbirds; Northern Flicker; Cordilleran Flycatcher; Say's Phoebe; Cassin's Kingbird; Plumbeous and Warbling Vireos; Northern Rough-winged, Violet-green, and Barn Swallows; Virginia's and MacGillivray's Warblers; Common Yellowthroat; Yellow Warbler; Spotted and Canyon Towhees; Hepatic and Western Tanagers; Black-headed Grosbeak; Common Grackle; Brown-headed Cowbird; Bullock's Oriole; and Lesser Goldfinch.

During migration, look for Olive-sided Flycatcher, Tree Swallow, Wilson's Warbler, Chipping and Lincoln's Sparrows, and Brewer's Blackbird. The area often attracts rarities in addition to these regularly occurring migrants.

Species present during the winter include Bald Eagle, Red-naped Sapsucker, American Dipper, Cedar Waxwing, Yellow-rumped Warbler, Dark-eyed Junco, and American Goldfinch.

DIRECTIONS

From Santa Fe: From the intersection of I-25 (Exit 282) and St. Francis Drive (US 84/285), take I-25 northbound (heading southeast) 16.5 miles to Exit 299 (Glorieta). Cross the overpass and take NM 50 east approximately 6 miles to the Village of Pecos. Turn north onto NM 63 and travel approximately 1.7 miles north to the left turn for Monastery Lake.

From Las Vegas, New Mexico: From I-25 Exit 345 (NM 104), travel south on I-25 approximately 38 miles to Exit 307 (Rowe). Take NM 63 north (through the Village of Pecos) approximately 7.7 miles to the left turn to Monastery Lake.

PARKING

There is a large dirt parking lot.

FEES

None

SPECIAL CONSIDERATIONS AND HAZARDS

See chapter 2 for further safety guidelines.

- Rattlesnakes: Rattlesnakes are always a possibility. Keep your eyes and ears alert, particularly early on summer mornings.
- Poison ivy: It is fairly abundant in this area.

FACILITIES

- Accessibility: Trails are fairly level, although narrow.
- Restrooms: Adjacent to the parking lot

- Water: None available
- Picnic tables: Located near the lake
- Pets: New Mexico Department of Game and Fish requires dogs to be on leash.

CAMPING

No on-site camping. Nearest camping is in Santa Fe National Forest sites on NM 63.

GAS, FOOD, AND LODGING

Gas and food are available in the Village of Pecos. Bed-and-breakfasts, guest ranches, and cabins are available at various locations in the general area of Pecos. There are motels and hotels in Las Vegas and Santa Fe, both about 30 minutes away in opposite directions on I-25.

Along NM 63

Description

As you drive north from Monastery Lake, NM 63 follows the Pecos River as it alternately winds through narrow canyons bounded by tall cliffs and open valleys. With the exception of the Santa Fe National Forest and New Mexico Department of Game and Fish areas, all of the areas are private property where access to the river is restricted. In addition to montane species, the fast currents and riffles at many places along the Pecos River provide excellent habitat for American Dipper. The Pecos River is stocked by New Mexico Department of Game and Fish, making the area popular with anglers. Campgrounds and fishing access areas are often crowded during summer weekends.

Two large wildfires in early summer 2013, followed by a long season of monsoonal rains, affected the habitat and access to some of the sites along the upper Pecos River. Wildlife managers expect all sites to be open by 2015.

This section includes various locations worth exploring for birds between Monastery Lake and Jacks Creek. There are separate site descriptions for both Monastery Lake and Jacks Creek.

Lower Dalton Day-Use (on the left as you travel up the canyon) and Upper Dalton Fishing Area (across the highway on the right as you travel up the canyon), both Santa Fe National Forest properties (7,200 feet): Located 5.8 miles north of Monastery Lake. The day-use area, nestled under the trees, provides better birding.

American Dipper (photo by Bernard R. Foy)

Bert Clancy Fishing Area, owned and managed by the New Mexico Department of Game and Fish (7,650 feet): Located on the right 6.7 miles past
the Dalton area. It provides access to the Pecos River in a ponderosa pine
habitat.

Holy Ghost Campground (8,150 feet): Located at the end of Forest Road
122 in a ponderosa pine–lined canyon along Holy Ghost Creek, this campground provides excellent montane riparian birding. If you do not intend to
camp or picnic, park at the trailhead lot. Bird in the campground and along
the creek. It can be worthwhile to continue up the trail to Spirit Lake for a
while. A bridge crosses the creek a short distance past the camping area. The
trail continues following the creek and passes through montane meadow, riparian habitat, and mixed conifer habitat.

Terrero area (7,691 feet): There are two areas to explore. County Road
122/Holy Ghost Canyon Road, 0.4 mile past Bert Clancy Fishing Area, veers
to the left. Almost immediately, a road on the right leads to the Game and
Fish Department's Terrero Campground in an open area at the foot of a cliff.
Bird the wooded edges of the campground. When you return to NM 63, stop

at the turnout across from an old bridge. American Dipper often nests under this bridge and frequently can be seen in this part of the river. Continue to the Terrero General Store to enjoy the hummingbird feeders that attract Broad-tailed, Rufous, and Calliope Hummingbirds during the summer.

As you continue up NM 63, the road becomes narrower with winding curves in many places. After about 5 miles, the Pecos River flows through an area called the Terrero Box, a canyon below the level of the road.

Cowles Ponds and Campground (8,561 feet): Turn left off NM 63 and then immediately pull into the parking lot on your right. From here you can bird ponderosa habitat along the river and scan for American Dipper.

Jacks Creek: See separate site description.

A map showing these locations can be found here: http://www.pecosnew mexico.com/images/stories/maps/maps2013/FINAL-Pecos_map_2013.pdf

County: San Miguel

eBird Hotspots: Santa Fe NF—Holy Ghost Campground

Target Birds

Bald Eagle It can be seen along the Pecos River during the winter.

Broad-tailed, Rufous, and Calliope Hummingbirds Broad-tailed Hummingbird can be seen starting in April all along this route and is present through mid-September. Rufous and Calliope arrive in July and also are present through mid-September. The easiest location to see all three species is at the feeders outside the Terrero General Store.

Belted Kingfisher Listen for its rattling call as it flies along the river at any time of year.

Plumbeous and Warbling Vireos Both vireos can be seen along this route. They are often difficult to spot. Listen for their songs to locate them. The Plumbeous Vireo, found midlevel in pines, has an up-and-down whistled song, while the Warbling Vireo, found in cottonwoods and aspens near water, has a warbling song. They both can be seen between May and September.

Steller's Jay This crested jay of western montane woodlands can be found at any location year-round.

American Dipper It can be found all along the Pecos, where the river is fast flowing over riffles. It often nests under bridges or overhanging rocks. The best chance to see one is to go to an appropriate spot and wait.

Western Tanager This colorful tanager prefers conifer woodlands. It arrives in early May and can be present through September.

Evening Grosbeak While it is a year-round resident, it is more likely to be seen in June and August/September.

Other Birds

Species that can be seen at any time of year include Red-tailed Hawk; Great Horned and Flammulated Owls; Western Screech-Owl; Hairy Woodpecker; Steller's Jay; Black-billed Magpie; American Crow; Common Raven; Mountain Chickadee; Red-breasted, White-breasted, and Pygmy Nuthatches; Brown Creeper; and American Robin (April–October).

In the summer, look for Turkey Vulture; Cooper's Hawk; Band-tailed Pigeon; Northern Saw-whet Owl; Williamson's Sapsucker; Northern Flicker; Western Wood-Pewee; Cordilleran Flycatcher; Violet-green Swallow; Ruby-crowned Kinglet; Hermit Thrush; American Robin; Orange-crowned, Virginia's, and Yellow-rumped Warblers; Green-tailed Towhee; Dark-eyed Junco; Black-headed Grosbeak; and Pine Siskin.

During migration, you might see House Wren; Golden-crowned and Ruby-crowned Kinglets; Western Bluebird; MacGillivray's, Townsend's, and Wilson's Warblers (fall); Chipping and White-crowned Sparrows; and Cassin's Finch.

DIRECTIONS

From Santa Fe: From the intersection of I-25 (Exit 282) and St. Francis Drive (US 84/285), take I-25 northbound (heading southeast) 16.5 miles to Exit 299 (Glorieta). Cross the overpass and take NM 50 east approximately 6 miles to the Village of Pecos. Turn north onto NM 63 and travel approximately 7.5 miles north to the Dalton Day-Use and Fishing Areas.

From Las Vegas, New Mexico: At I-25 Exit 345 (NM 104), travel south on I-25 approximately 38 miles to Exit 307 (Rowe). Take NM 63 north (through the Village of Pecos) approximately 13.5 miles to the Dalton Day-Use and Fishing Areas.

PARKING

There are parking areas at each of these sites.

FEES

There are day-use and camping fees at Forest Service locations. NM Game and Fish properties require a GAIN permit and an HMAV, which are not available on-site. (See chapter 2 for information on obtaining the permit and validation.)

SPECIAL CONSIDERATIONS AND HAZARDS
See chapter 2 for further safety guidelines.

- Postfire hazards: Watch for rolling rocks, stump and root holes, washed-out trails and roads, debris flows, and falling trees.
- Flash floods: The potential for flash floods increases with exposed areas as a result of wildfires. Do not attempt to drive across an area with water on the road.
- Summer thunderstorms: Thunderstorms are expected during the summer months and can materialize without warning.
- Winter driving conditions: NM 63 can be snowpacked and icy during the winter. It is advisable to check the New Mexico Department of Transportation website (www.nmroads.com) or call the hotline (800–432–4269) for current road conditions during winter months.
- Bears: Black bears are possible at any time. Do not leave food unattended and be aware of safety precautions.
- Rattlesnakes: Prairie rattlesnakes can be present at elevations up to 9,500 feet. While not prevalent, they can be found at any site. Just be aware that they can be out there.
- Poison ivy and stinging nettle: While not abundant, both plants are present in some areas near streams.
- Cell phone service: There is no cell phone coverage north of Monastery Lake.

FACILITIES

- Accessibility: Most of the locations are fairly level. Fishing trails along the river may be uneven. Cowles has ADA accessible restrooms. Cowles Ponds has an ADA accessible path that allows access to both ponds.
- Restrooms: All areas have restrooms, except the Lower Dalton Day-Use Area.
- Water: Water is available at the Field Track Campground (below Bert Clancy Fishing Area) and at the Cowles Campground.
- Picnic tables: Lower Dalton Day-Use Area and Cowles Campground

- Pets: Dogs must be kept on leash.

In addition to the campgrounds mentioned previously, there are Forest Service and Game and Fish properties with camping along NM 63.

GAS, FOOD, AND LODGING
The closest gas station and restaurants are in the Village of Pecos. Some groceries can be purchased at the Terrero General Store. There are numerous options for lodging along NM 63.

Jacks Creek Campground
Description
Situated in an alpine meadow at almost 9,000 feet, this Santa Fe National Forest campground is surrounded by mixed conifer habitat and aspens.

As you drive north on NM 63, look for a turnout on your left about 1.25 miles from Cowles. Park at the turnout and carefully walk along the

Jacks Creek Campground

shoulder to the drainage on your left. This is an excellent area to look for montane riparian species. As you approach the campground, turn right onto the road that leads to the Wilderness Area parking and horse corrals. You can park near the wilderness information sign/Jacks Creek Trailhead. You can walk this loop and then walk over to the campground loop, which will provide opportunities for both montane meadow and mixed conifer species.

There are trails that climb into the Pecos Wilderness from Jacks Creek Trailhead; however, they are heavily used by those on horseback and therefore provide challenges for birding.

The campground opens in May and closes late December/early January, depending on the weather.

County: San Miguel

eBird Hotspots: Santa Fe NF—Jacks Creek Campground

Target Birds

Northern Goshawk This montane accipiter, while not guaranteed, has been seen at this location.

Williamson's Sapsucker While not prevalent, it nests in this area in the mixed conifer woodlands. It may be challenging to glimpse as it moves silently through the forest.

Ruby-crowned Kinglet It arrives in mid-May and can be found high in the conifers, where it moves rapidly through the tops of the trees. It can be present through the end of September.

Orange-crowned Warbler This drab warbler prefers dense deciduous understory. Look for it along Jacks Creek between mid-May and September.

Green-tailed Towhee It may be seen between the end of May and the end of September in brushy areas at the edges of the meadow.

Western Tanager It can be spotted in the upper areas of the trees. The male arrives first in early May, followed by the female. Toward the end of summer, the male is the first to depart, followed by females, and then the immatures.

Other Birds

Species that can be seen anytime the campground is open include Red-tailed Hawk, Hairy Woodpecker, Steller's Jay, American Crow, Common Raven, Mountain Chickadee, White-breasted Nuthatch, American Dipper (dependent on water levels in Jacks Creek), American Robin, and Red Crossbill.

During the summer months, you might find Turkey Vulture, Sharp-shinned and Cooper's Hawks, Broad-tailed Hummingbird, Northern Flicker, Western Wood-Pewee, Cordilleran Flycatcher, Warbling Vireo, Violet-green Swallow, Yellow-rumped Warbler, Dark-eyed Junco, Black-headed Grosbeak, Pine Siskin, and Lesser Goldfinch (late summer).

Species that can be seen during migration include Olive-sided (fall) and Hammond's Flycatchers, House Wren, Mountain Bluebird, Chipping and Lincoln's Sparrows, and Cassin's Finch.

DIRECTIONS

From Santa Fe: From the intersection of I-25 (Exit 282) and St. Francis Drive (US 84/285), take I-25 northbound (heading southeast) 16.5 miles to Exit 299 (Glorieta). Cross the overpass and take NM 50 east approximately 6 miles to the Village of Pecos. Turn north onto NM 63 and travel about 21.7 miles to the end of the road at Jacks Creek Campground.

From Las Vegas, New Mexico: At I-25 Exit 345 (NM 104), travel south on I-25 approximately 38 miles to Exit 307 (Rowe). Take NM 63 north (through the Village of Pecos) approximately 27.7 miles to Jacks Creek Campground.

PARKING

There is day-use parking at the trailhead.

FEES

Camping, day-use, and trailhead parking fees. Fees are discounted for federal pass holders.

SPECIAL CONSIDERATIONS AND HAZARDS

See chapter 2 for further safety guidelines.

- Black bears: Black bears are possible at any time. Do not leave food unattended.
- Cougars: Cougars are found in the national forest and wilderness area, most frequently at dawn and dusk.
- Summer thunderstorms: Thunderstorms can materialize without warning. Keep an eye on the sky to avoid getting caught on the trail during a storm.
- Trailhead parking: The Santa Fe National Forest website suggests not leaving valuables in your vehicle.

FACILITIES

- Accessibility: The area around the campgrounds and trailhead is level.
- Restrooms: There are several accessible restrooms.
- Water: Water is available in the campground.
- Picnic tables: Day-use visitors can use tables in the campground area.
- Pets: Santa Fe National Forest rules require that dogs be on leash.

CAMPING

Campground on-site

GAS, FOOD, AND LODGING

The closest gas and restaurants are in the Village of Pecos. Groceries are available at Terrero General Store. There is a guest ranch at Cowles and other similar lodging along NM 63.

Villanueva State Park

Description

Villanueva State Park, located along the Pecos River at 6,165 feet, sits in a transition zone between the Rocky Mountains and the Great Plains, providing varied birding habitats. Sandstone cliffs rise up on both sides of the river as it flows through the state park. This is a very popular camping, fishing, and river tubing spot on the weekends. For the best birding, visit midweek during the summer season.

If you are arriving from the north, you will follow the Pecos River through historic land grant towns and farms. This is all private property that does not provide many opportunities for birding. If you are arriving from the south in May and June, drive with your windows open starting about 1 mile north of I-40 to listen for breeding Cassin's Sparrow. The area all along NM 3 is private property. Do not leave the road shoulder.

Before driving into the state park, travel south less than 1 mile through the village and across the bridge over the Pecos (if you are arriving from the south, stop just before arriving at the bridge). Immediately after crossing the bridge, there is an unmarked County Road (B29B) with a turnout area along the east side of NM 3. Park in the turnout and walk along B29B (May through July) to get a good view of the swallows that nest under the bridge and hunt

Pecos River, Villanueva State Park

for insects along the river. You might be lucky and see Cliff Swallow nestlings peeking out of their cone-shaped nests. Survey the riparian area along the river. Next, walk back up to NM 3 and cross to a wide area on the shoulder on the west side of the highway, which will give you a good viewpoint of the riparian area. This is a good location to search for Eastern Phoebe.

After entering the park, stop and park near the pay station. This is good vicinity for birding along the edge of the river. Swallows and Say's Phoebe often nest in the cliffs just west of the restrooms. There is a campsite at the base of the cliffs, so please respect the privacy of the campers if the site is occupied. Western Wood-Pewee and American Robin often nest in the trees in this area.

Next, drive or walk farther into the campground. There is parking in front of the state park office (where there are displays as well as a bird and herp list). From this location walk past the restrooms to the footbridge that crosses the river. This is an ideal vantage point for Black Phoebe, as well as Rock and Canyon Wrens. This is also the start of the Viewpoint Loop Trail.

Finally, drive or walk to the end of the campground. You can park in the day-use area. Look for birds in the trees near the road and the riparian area along the river north of the campsites.

Villanueva State Park is open for day-use visitors from 7:00 a.m. to 9:00 p.m., April 1–September 30, and 7:00 a.m. to 7:00 p.m., October 1–March 30.

County: San Miguel

eBird Hotspots: Villanueva SP and Villanueva Bridge over Pecos River

Target Birds

Western Wood-Pewee Arriving in May, they are present through the end of July. If you watch their behavior, you may see the adults taking food to nestlings.

Black Phoebe It nests along the Pecos River in the campground. A good location to search for it is on the footbridge leading to the Overlook Trail.

Eastern Phoebe This is a reliable nesting location for this species, which is seen in only a limited number of locations in New Mexico north of I-40. A good place to look for it is north of the bridge over the Pecos River just south of the village.

Violet-green and Cliff Swallows Cliff Swallows nest under the Villanueva Bridge, as well as in cliff niches near the entrance station in the campground. Violet-green Swallows often help feed Cliff Swallow chicks and then take over the nest along the cliffs when it becomes empty.

Cedar Waxwing This nomadic species might be seen at any time of year as it searches for fruits and berries. In winter it might be found eating Russian olive fruit.

Yellow-breasted Chat It is a regular summer resident and breeder along the Pecos River both in the state park and along the river south of the village.

Hepatic Tanager This is good habitat for Hepatic Tanager, and it has been seen at this location, although it is not guaranteed.

Other Birds

Species that can be seen at any time of year include Great Blue Heron; Cooper's and Red-tailed Hawks; Rock Pigeon; Eurasian Collared-Dove; White-winged Dove; Great Horned Owl; Ladder-backed, Downy, and Hairy Woodpeckers; Northern Flicker; American Kestrel; Pinyon Jay; Western Scrub-Jay; Black-billed Magpie; American Crow; Common Raven; Mountain Chickadee; Juniper Titmouse; Bushtit; White-breasted Nuthatch; Brown Creeper; Canyon Wren; European Starling; Canyon and Spotted Towhees; Red-winged Blackbird; House Finch; and House Sparrow.

In the summer, look for Turkey Vulture, Swainson's Hawk, Mourning Dove, Common Nighthawk, Black-chinned and Broad-tailed Hummingbirds, Say's Phoebe, Ash-throated Flycatcher, Cassin's and Western Kingbirds, Northern Rough-winged and Barn Swallows, Rock Wren, Yellow Warbler, Chipping Sparrow, Black-headed Grosbeak, Brown-headed Cowbird, and Bullock's Oriole.

During migration you might see Chimney Swift (unpredictable); Olive-sided Flycatcher; Tree Swallow; Gray Catbird; Townsend's (fall), Virginia's, MacGillivray's, and Yellow-rumped Warblers; Western Tanager; and Lazuli Bunting.

Birds that winter in the area include Common Merganser, Ruby-crowned Kinglet, Western Bluebird, Townsend's Solitaire, Hermit Thrush, White-crowned Sparrow, and Dark-eyed Junco.

DIRECTIONS

From Santa Fe: From the intersection of I-25 (Exit 282) and St. Francis Drive (US 84/285), take I-25 northbound (heading southeast) approximately 40 miles to Villanueva Exit 323. Turn right onto NM 3 and travel approximately 12 miles south to the small community of Villanueva. The state park signs in town will direct you to turn east and travel about 2 miles to the state park entrance.

From Las Vegas, New Mexico: From I-25 Exit 345 (NM 104), travel south on I-25 approximately 23 miles to Villanueva Exit 323. Turn south onto NM 3 and travel approximately 12 miles south to the small community of Villanueva. The state park signs in town will direct you to turn east and travel about 2 miles to the state park entrance.

From Albuquerque: From the intersection of I-40 and I-25, travel approximately 71 miles east on I-40 to Encino/Villanueva Exit 230 (about 13 miles east of Clines Corners). Turn north onto NM 3 and travel approximately 20 miles to the community of Villanueva. After crossing the bridge over the Pecos River, travel east 2 miles to the park entrance.

PARKING

Parking for day use is available near the pay station, in front of the state park office, and in the day-use picnic area.

FEES

Day-use fee or state parks annual day-use permit

SPECIAL CONSIDERATIONS AND HAZARDS

See chapter 2 for further safety guidelines.

- Rattlesnakes: Two species of rattlesnakes live in the area. They are often encountered in this park near sawn log barriers and at the base of trees.
- Black bears: Black bears are possible at this location. Be sure not to leave food unattended. When you are finished picnicking, secure food in your vehicle before birding.
- Mosquitoes: Mosquitoes are possible in the summer near the river.

FACILITIES

- Accessibility: Some campsites and restrooms are fully accessible. Birding in the campground area is on level ground. The Viewpoint Loop Trail is a strenuous, uneven, and rocky trail. The 0.28-mile River Trail is fairly level but uneven in spots.
- Restrooms: There are accessible restrooms throughout the park.
- Water: Water is available in camping areas.
- Picnic tables: There is a separate day-use area with picnic tables at the north end.
- Pets: Park rules require that dogs be on leash.

CAMPING

Camping is available in the park.

GAS, FOOD, AND LODGING

Gas and convenience store/snack bars are located north in Romeroville (I-25) and south at Clines Corners (I-40). There are no restaurants in the general area. The closest is in Las Vegas, New Mexico. There is a general store in the community of Villanueva. Closest lodging is in Las Vegas or Santa Fe.

▶ CHAPTER 10

Along I-25 North of Santa Fe

General Overview

The four sites in this chapter are located off I-25 between Las Vegas and Maxwell, including two national wildlife refuges, one New Mexico Department of Game and Fish Wildlife management area, and a New Mexico state park.

I-25 roughly follows the National Historic Santa Fe Trail and Scenic Byway. As you drive northbound from Santa Fe, the route goes through Apache Canyon, famous for the Civil War Battle of Glorieta Pass. Continuing east from the Pecos area, I-25 drops down onto the edge of the Great Plains. As you travel north, the Sangre de Cristo Mountains are a backdrop to the northwest. Look for pronghorn in fields and ranches that border the highway. Located on the Central Flyway, these sites provide habitat for migrating raptors, waterfowl, and shorebirds in addition to songbirds.

All sites are open year-round and provide excellent winter birding. Because driving conditions may be difficult during or following snowstorms, it is advisable to check the New Mexico Department of Transportation website (www.nmroads.com) or call the hotline (800–432–4269) for current road conditions during winter months.

Las Vegas National Wildlife Refuge
Description

The refuge is located just south of the Santa Fe Trail Scenic Byway at the convergence of two life zones: Great Plains prairie grasslands to the east and the Rocky Mountains to the west form an ecotone, which is also influenced by the Chihuahuan Desert to the south. Within this ecotone are a number of habitat types, including piñon-juniper woodlands, middle-elevation riparian areas, wetlands and ponds, and plains-mesa grassland. The refuge's varied habitats and location on the Central Flyway attract a large variety of migrating and resident bird species, including 50 that nest on the refuge grounds. The refuge has been designated an Important Bird Area by Audubon New Mexico.

Begin birding as you turn south from NM 104 and drive the 4.5 miles along NM 281 to the refuge headquarters. The highway passes through farmland and provides opportunities to see raptors, sparrows, and other grassland birds, but the land on either side of the road is private property. Be sure to pull well off the highway when stopping to look for birds, being careful not to block property access routes. NM 281 turns east for about 1 mile, in an area where the homes are closer to the road and hedgerows and windbreaks have been planted. This area attracts riparian species. Shortly after curving south again, the road enters the refuge property.

At this point the 8.5-mile auto tour route (NM 281 and CR C22C) begins and encompasses the central portion of the refuge. It is open during daylight hours and offers opportunities for wildlife watching year-round.

Melton Pond north of the headquarters area can be viewed with a spotting scope from the parking lot. Meadowlark Trail, beginning at the refuge gate, is a 0.5-mile self-guided nature trail that traverses shortgrass prairie, past an irrigation canal and wetland and along the hedgerow. The 0.53-mile accessible, interpretive trail begins at the refuge headquarters and circles the Visitor Center.

If you visit on a Sunday in November, follow the Fall Flight Wildlife Drive into the normally closed portions of the refuge to the east. This drive approaches three more lakes (including Goose Island Lake), providing closer views of migratory waterfowl and wintering raptors. In addition to the open wildlife drive, there are special educational programs in the Visitor Center at these times.

After exiting the headquarters area, turn left and travel almost 1 mile along NM 281 to the parking area and overlook for Crane Lake. Before heading to the viewing platform, take a few minutes to inspect the windbreak on the south edge of the parking area. While it is best to bring a spotting scope, the platform has two mounted viewing scopes on the observation deck. During the winter months, Crane Lake accommodates a variety of wintering waterfowl, which at times includes rare species.

Continue south on NM 281 past Middle Marsh Ponds, where the road

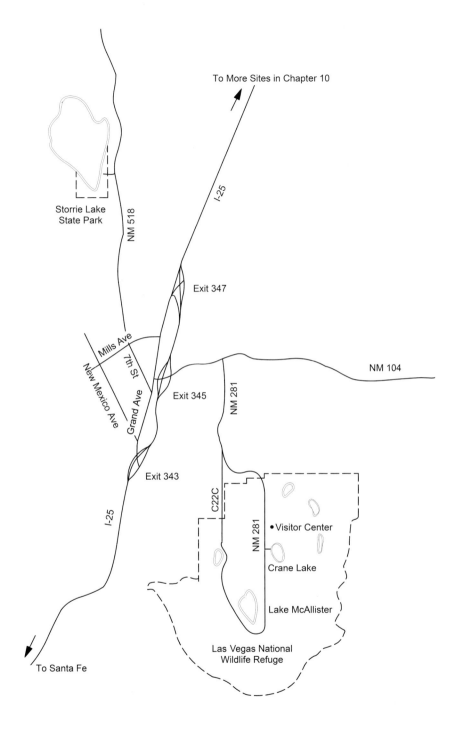

To More Sites in Chapter 10

Storrie Lake
State Park

NM 518

I-25

Exit 347

Mills Ave

7th St

New Mexico Ave

Grand Ave

Exit 345

NM 281

NM 104

Exit 343

C22C

NM 281

•Visitor Center

Crane Lake

Lake McAllister

Las Vegas National
Wildlife Refuge

I-25

To Santa Fe

N

Map 12. Las Vegas National Wildlife Refuge and Storrie Lake State Park

turns west and becomes County Road C22C. The road from this point may be muddy and difficult to drive during extremely rainy weather or in winter. At the southern end of the refuge, the McAllister Lake Wildlife Management Area, managed by the New Mexico Department of Game and Fish (NMDGF), once a prime birding area and eBird Hotspot, was closed in 2011 to vehicle access and fishing because of low water levels as a result of the continued drought. Check the NMDGF website regarding its possible reopening status. The parking area for the Gallinas Canyon Nature Trail is located just north of where C22C makes a right turn and heads north. The 2-mile (round-trip) steep trail transitions from grasslands into a canyon with piñon-juniper habitat and gives the opportunity to see species such as Canyon and Rock Wrens. The county road that borders the west side of the refuge passes through rangeland-grassland habitat and is dotted with farms.

County: San Miguel

eBird Hotspots: Las Vegas NWR; Las Vegas NWR—Crane Lake; Las Vegas NWR—Goose Island Lake; Las Vegas NWR—HQ Hedgerow; Las Vegas NWR—Gallinas Canyon; and McAllister Lake WMA

website: http://www.fws.gov/southwest/refuges/newmex/lasvegas/index.html

Facebook: https://www.facebook.com/FLVNWR

Target Birds

Snow and Ross's Geese Flocks of Snow and Ross's Geese winter at the refuge, often huddled in a group near Crane Lake. There may be an opportunity to see the size and bill differences in these two species of geese if they are disturbed by a hawk or an eagle.

Tundra Swan One or two, although sometimes as many as five or six, Tundra Swans usually take up residence at the refuge between November and mid-March.

Bald Eagle A frequent winter visitor, it can be seen patrolling over Crane Lake.

Ferruginous Hawk It arrives in the area by early November and is present through early March. Both light and dark morphs are possible.

Rough-legged Hawk One or two are normally present between early November and early March, although they have been seen as late as mid-March.

Prairie Falcon It can be seen year-round at the refuge. You might see it pursuing prey close to the ground.

Birders at Crane Lake

Cassin's and Western Kingbirds The breeding range of these species overlaps at the refuge, with both species present. Due to feather wear, the tail feathers are not always a good way to differentiate these species. More reliable is the whiter chin, contrasting with a darker head that can be observed in the Cassin's.

Northern and Loggerhead Shrikes While Loggerhead Shrike can be seen year-round, Northern Shrike normally is present only from mid-November through January. Careful observation is necessary to discern these two shrikes.

Cassin's Sparrow A few breed at the refuge and are most easily detected near McAllister Lake. While they arrive in April, they are most easily seen in late June and through July when they are singing and "skylarking."

Lark Bunting The refuge is on its migration path. Although it sometimes makes an appearance in May, it is more frequently seen during fall migration during August and early September.

Other Birds

The refuge hosts large numbers of waterfowl, both during the winter and as these species migrate north and south. Regularly occurring species include

Cackling Goose, Gadwall, American Wigeon, Northern Shoveler, Northern Pintail, Green-winged Teal, Lesser Scaup, Canvasback, Redhead, Ring-necked Duck, Bufflehead, Common Goldeneye, and Common and Hooded Mergansers. Sandhill Crane and Ring-billed Gull also winter at the refuge or stop there during migration.

In addition to the raptors highlighted as target birds, a number of other raptors winter at the refuge, including Northern Harrier, Sharp-shinned Hawk, and Merlin. Hairy Woodpecker winters here. Wintering passerines include Marsh Wren; Western and Mountain Bluebirds; Townsend's Solitaire (irregular); American Pipit; Chestnut-collared Longspur; American Tree, Song, and White-crowned Sparrows; Dark-eyed Junco; and American Gold-finch.

Migrating waterfowl include Blue-winged and Cinnamon Teal. Eared, Western, and Clark's Grebes use the refuge as a staging area after breeding. Double-crested Cormorant and American White Pelican can be seen in the fall. White-faced Ibis stops in both spring and fall. A plethora of shorebirds stop at the refuge; however, they can be difficult to see. These include Spotted Sandpiper; Lesser (fall) and Greater Yellowlegs; Willet; Long-billed Curlew;

Rough-legged Hawk (photo by Joe Schelling)

Marbled Godwit; Western, Least, and Baird's (fall) Sandpipers; Long-billed Dowitcher; and Wilson's Phalarope. Franklin's Gull can be seen during spring migration. Rufous and Calliope Hummingbirds are seen in the fall.

Passerines possible during migration include Tree and Violet-green Swallows; House Wren; Orange-crowned, Virginia's, MacGillivray's, Yellow, Yellow-rumped, and Wilson's (fall) Warblers; Clay-colored (fall), Savannah, and Lincoln's Sparrows; Western Tanager; and Lazuli Bunting (fall).

During the summer, look for Turkey Vulture; Swainson's Hawk; Killdeer; Common Nighthawk; Common Poorwill; Black-chinned and Broad-tailed Hummingbirds; Western Wood-Pewee; Northern Rough-winged and Barn Swallows; Rock Wren; Northern Mockingbird; Chipping, Vesper, and Lark Sparrows; Blue Grosbeak; Dickcissel; Yellow-headed Blackbird; Common Grackle; Brown-headed Cowbird; Bullock's Oriole; and Lesser Goldfinch.

Year-round birds include Canada Goose, Ruddy Duck (February–November), Pied-billed Grebe (February–November), Great Blue Heron, Cooper's and Red-tailed Hawks, Rock Pigeon, Eurasian Collared-Dove, Mourning Dove, Greater Roadrunner, Barn and Great Horned Owls, Western Screech-Owl (normally heard rather than seen), Ladder-backed Woodpecker, Northern Flicker, American Kestrel, Say's Phoebe, Western Scrub-Jay, Black-billed Magpie, American Crow, Common Raven, Horned Lark, Bewick's Wren, American Robin, Red-winged Blackbird, Western Meadowlark, Great-tailed Grackle, and House Finch.

DIRECTIONS

In Las Vegas, New Mexico, from the intersection of I-25 (Exit 345) and NM 104 (NM 65/East University Avenue), travel east on NM 104 approximately 1.2 miles. Turn right (south) on NM 281 and travel about 4.5 miles to the refuge headquarters.

PARKING

There is a large parking area adjacent to the headquarters building as well as a smaller parking area next to the road that can be used when the headquarters and gate are closed. There is also parking at Crane Lake and at the Gallinas Canyon Nature Trail.

FEES
None

SPECIAL CONSIDERATIONS AND HAZARDS

See chapter 2 for further safety guidelines.

- Rattlesnakes: Two species of rattlesnakes live on the refuge. A visitor might encounter one hiking on one of the trails or on the roads.
- Resource protection: All plants, wildlife, and cultural features on the refuge are protected, and it is illegal to remove them.
- Muddy roads: After a snowfall or heavy rain, some refuge roads may become impassable.
- Winter driving conditions: The highways leading to the refuge may be icy, snowpacked, or closed during the winter.

FACILITIES

- Accessibility: The headquarters parking area, building, and Meadowlark Trail are fully accessible. The Gallinas Canyon Nature Trail is steep and uneven at the westernmost half-mile section.
- Restrooms: Located in the headquarters building. A vault toilet is located in the parking area when the building is closed.
- Water: Drinking fountain available in the headquarters building (when open)
- Picnic tables: Adjacent to the headquarters building
- Pets: Refuge rules require that pets be on leash.

CAMPING

Closest camping is at Storrie Lake State Park just north of Las Vegas.

GAS, FOOD, AND LODGING

There are abundant options for gas, food, and lodging in Las Vegas.

Storrie Lake State Park

Description

Located 4 miles north of Las Vegas, this state park has been a popular campground for fishing and windsurfing. Nestled against the Sangre de Cristo Mountains, it is situated on the edge of the southern Great Plains in a plains-mesa grassland habitat. The lake, formed by a dam on the Gallinas River, is an important source of water for irrigation and supports the seasonal ponds at the Las Vegas NWR. During years of drought, the water level may drop quite low. Water levels can be very dynamic, rising and falling

Bald Eagle on Storrie Lake

depending on monsoonal storm activity (July–September). Because of the slope of the banks, lake size can increase substantially with a single monsoonal storm, creating large areas of shallow water surrounded by emergent vegetation.

The marshes on the western half of the lake support the most birds from September through the winter. The lake occasionally freezes, which can afford great views of Bald Eagles hunting and occasionally eating waterfowl trapped on the ice.

In addition to exploring the edges of the lake, drive along the campground road toward the RV camping area on the south. Across the road is a band of riparian habitat that borders a tributary stream. The crumbling bridge that once crossed the creek is impassable. Explore the clumps of trees and grassland bordering the park roads.

About 0.25 mile south of the park on the west side of NM 518 is a private farm pond that attracts waterfowl, particularly during the winter months. It can be viewed from the side of the road.

County: San Miguel

eBird Hotspots: Storrie Lake SP and Mallete's Pond (from Hwy 518)
website: http://www.emnrd.state.nm.us/SPD/storrielakestatepark.html

Target Birds

Redhead A fair number of Redheads can be seen during the winter months on the lake, as well as the pond south of the state park.

Common Merganser Sizable numbers of this diving duck can be seen between November and February.

Western and Clark's Grebes Both are present April through November.

American White Pelican Large numbers can be seen at times during both spring and fall migration.

Great Blue Heron Look for Great Blue Heron around the edges of the lake at any time of year.

Ring-billed Gull It winters at the lake between November and March.

Cassin's and Western Kingbirds The park is located where their ranges overlap, so both species of kingbird can be seen between May through early September.

Mountain Bluebird While it nests not too far west of the lake, it can be seen in the grassy areas of the park after the breeding period.

Other Birds

Species generally present during the winter include Gadwall, Green-winged Teal (irregular), Canvasback, Ring-necked Duck, Bufflehead (irregular), Common Goldeneye, Bald Eagle, Belted Kingfisher, Northern Flicker, Song Sparrow, and Dark-eyed Junco.

Storrie Lake is located on the Central Flyway and attracts a number of migrating species, including American Wigeon; Northern Shoveler; Blue-winged Teal; Northern Pintail; Eared Grebe; American White Pelican; Osprey; Northern Harrier; American Avocet; Long-billed Curlew; Western, White-rumped, and Baird's Sandpipers (all irregular); Northern Rough-winged, Tree, and Violet-green Swallows; House and Marsh Wrens; Orange-crowned, MacGillivray's, Yellow, Yellow-rumped, and Wilson's Warblers (fall only); Green-tailed Towhee (fall); Chipping and Savannah Sparrows; Blue Grosbeak; and Yellow-headed Blackbird.

During the summer months, look for Double-crested Cormorant, Turkey Vulture, Swainson's Hawk, Peregrine Falcon, Killdeer, Spotted Sandpiper, Broad-tailed Hummingbird, Western Wood-Pewee, Say's Phoebe, Ash-throated Flycatcher, Barn and Cliff Swallows, Northern Mockingbird, Lark Sparrow, Brewer's Blackbird, Common Grackle, Pine Siskin, and Lesser Goldfinch.

Year-round species include Canada Goose; Mallard; Ruddy Duck; Red-tailed Hawk; American Coot; Mourning Dove; American Kestrel; Lewis's (irregular), Downy, and Hairy Woodpeckers; Northern Flicker; Black-billed Magpie; American Crow; Common Raven; Horned Lark; Black-capped Chickadee; European Starling; Red-winged Blackbird; Great-tailed Grackle; and House Finch.

DIRECTIONS

In Las Vegas, New Mexico (off I-25), from the intersection of Grand Avenue (US 85) and 7th Street (NM 518), travel northwest through the city on NM 518 approximately 4.7 miles to the park entrance.

PARKING

Scattered throughout the park

FEES

Day-use fee or state park annual day-use pass

SPECIAL CONSIDERATIONS AND HAZARDS

See chapter 2 for further safety guidelines.

- Rattlesnakes: Rattlesnakes are possible in brushy areas outside campgrounds.
- Vehicular traffic: When walking along the roads to look for birds, be aware of traffic, particularly during the summer months.
- High fire danger: During dry months, be cognizant of fire restrictions.
- Winter: During winter months, the lake may be frozen and there may be snow on the ground. NM 518 may be snowpacked and icy.

FACILITIES

- Accessibility: Most of the park is fairly level. The only accessible restrooms are in the group reservation area.
- Restrooms: Located in each area of the park. Only vault toilets are open from October 1 through May 14.

- Water: Available between May 15 and September 30 or from businesses near Grand Avenue and Mills Avenue in Las Vegas·
- Picnic tables: Picnic shelters are located in primitive camping areas.
- Pets: Park rules require that dogs be on leash.

CAMPING

Improved and unimproved campsites are available on-site (fee).

GAS, FOOD, AND LODGING

Gas and food are available at the intersection of Grand Avenue and Mills Avenue in Las Vegas. Lodging is available along Grand Avenue.

Springer Lake Wildlife Area

Description

The lake, located approximately 5 miles west of the town of Springer, is stocked and maintained by NMGDF for fishing. It attracts seasonal waterfowl, migrating shorebirds, and grassland species. Begin birding as you drive west from Springer. After about 2 miles, the pavement ends and you will travel along a well-maintained dirt road to the lake. There is a pond on the west side of a correctional facility property surrounded by a stand of large trees. This area can be good for birding, which you can accomplish from the side of the road. Both from this location and as you proceed to the lake, scan the grassy areas for Horned Lark, Western Meadowlark, and other plains-mesa grassland species.

When you arrive at the lake, take the first entrance on the left and proceed to the nearby parking area. If drought has lowered lake levels, you can walk along the south side of the lake. This provides the best vantage point for viewing shorebirds that tend to congregate on the mudflats along the west side of the lake during migration. It is advantageous to use a scope to enable viewing from a distance so as not to disturb their feeding.

When you have finished birding along the south side of the lake, return to the road, turn left, and proceed across the dam. You can drive slowly and scan both sides of the levee for birds, but vehicles are prohibited from stopping on the dam. It is not advisable to walk across the bridge, because there are no places to step out of the way of an approaching vehicle. The road then curves west, where there is a turnout with adequate space to park and set up a scope. This will allow you to bird the north side of the lake.

Springer Lake

If you are not returning to Springer, consider leaving the lake area by continuing north through the grasslands and farming area approximately 2 miles to NM 58. This route will provide additional opportunities for viewing species that prefer this habitat. When you reach NM 58, you can turn left to travel to Cimarron Canyon State Park, or right to I-25.

County: Colfax

eBird Hotspots: Springer Lake

Web site (with map): http://www.wildlife.state.nm.us/conservation/wildlife_management_areas/documents/SpringerLake.pdf

Target Birds

Common Merganser It can be found on the lake, sometimes in large numbers, from the end of October through early April.

Ruddy Duck Springer Lake attracts sizable groups of this diving duck from the end of February through the end of November.

Bald Eagle The lake has an abundant supply of fish year-round, and Bald

Eagle now has been recorded nesting here. Some individuals winter here.

Least and Baird's Sandpipers These are the most prevalent sandpipers on the mudflats on the northwestern part of the lake, irregularly during spring migration and in larger numbers from August through mid-October.

Franklin's Gull During April and May, this is a good location for viewing migrating Franklin's Gull.

Horned Lark It is prevalent year-round in the grasslands surrounding the lake.

Western Meadowlark Look for it on the fence line along NM 468 and the grassy areas near the lake.

Other Birds

Species regularly occurring during the winter include Canvasback, Redhead, Common Goldeneye, Common Merganser, Northern Harrier, Ferruginous and Rough-legged Hawks, Chestnut-collared Longspur, American Kestrel, Merlin, and Prairie Falcon. Irregularly occurring winter species include Greater White-fronted, Snow, Ross's, and Cackling Geese; Red-breasted Merganser; Red-throated, Pacific, and Common Loons; Great Horned Owl; McCown's Longspur; and American Tree Sparrow.

During migration, look for Wood and Ring-necked Ducks; Gadwall; American Wigeon; Cinnamon and Green-winged Teal; Northern Shoveler; Northern Pintail; Lesser Scaup; Bufflehead; Hooded Merganser; Pied-billed (fall), Horned, and Eared Grebes; Double-crested Cormorant; American White Pelican (fall); White-faced Ibis; American Coot; Sandhill Crane; American Avocet; Spotted Sandpiper; Greater and Lesser Yellowlegs; Long-billed Curlew; Sanderling; Semipalmated (irregular), White-rumped (irregular, May), Western, and Stilt Sandpipers; Long-billed Dowitcher; Wilson's Phalarope; Sabine's (fall), California (irregular), and Ring-billed Gulls; Black (irregular) and Forster's (fall) Terns; Northern Flicker; Violet-green Swallow; Yellow-rumped Warbler; Chipping (fall), Vesper (irregular), Brewer's, Savannah, and White-crowned Sparrows; Dark-eyed Junco; and Yellow-headed and Brewer's Blackbirds.

Species found during the summer include Western and Clark's Grebes (larger numbers during migration); Turkey Vulture; Swainson's Hawk; Killdeer; Western Wood-Pewee; Say's Phoebe (March–September); Western, Cassin's, and Eastern Kingbirds; Barn and Cliff Swallows; Northern Mockingbird; Common Yellowthroat; Lark Sparrow; Blue Grosbeak; Common and Great-tailed Grackles; Bullock's Oriole; House Finch; and Lesser and American Goldfinches.

Year-round species include Canada Goose, Mallard, Great Blue Heron, Red-tailed Hawk, Mourning Dove (usually absent in winter), Black-billed Magpie, Common Raven, European Starling, and Red-winged Blackbird.

DIRECTIONS

From Las Vegas, New Mexico: From the intersection of I-25 and NM 104, travel north on I-25 approximately 66 miles to Exit 412 (Springer). At the exit ramp, turn right and travel through town about 2.3 miles on Maxwell Avenue, which becomes Railroad Avenue. Turn left on NM 468, traveling approximately 2.3 miles to the end of the pavement. Continue on gravel road C 17 about 2.4 miles to the lake.

From Cimarron Canyon State Park (3 miles east of Eagle Nest): At the eastern park boundary on US 64, travel east 13.5 miles to the town of Cimarron. At the junction of US 64 and NM 58 in Cimarron, turn right and travel on NM 58 approximately 16 miles to C 17 (gravel). Turn right and travel approximately 1 mile to the north end of Springer Lake.

PARKING

Two parking areas: larger one near boat ramp; small turnout on north side of lake

FEES

None

SPECIAL CONSIDERATIONS AND HAZARDS

See chapter 2 for further safety guidelines.
- Driving cautions: The road over the dam is narrow with potholes. Stopping or parking on the dam or boat ramp is not allowed. Stay on established roads.
- Rattlesnakes: Rattlesnakes are possible during warm weather, particularly in the early morning and evening.

FACILITIES

- Accessibility: While generally level, the terrain around the lake is uneven.

- Restrooms: There are accessible restrooms on both sides of the lake.
- Water: None available
- Picnic tables: There are picnic tables near a boat ramp located in an unmaintained, weedy area.
- Pets: Dogs should be kept on leash.

CAMPING

Primitive camping is available on-site.

GAS, FOOD, AND LODGING

Gas and food are available in Springer and at the travel center at the Cimarron turnoff, I-25 Exit 419. There is a motel in Springer.

Maxwell National Wildlife Refuge

Description

Located in shortgrass prairie just east of the Sangre de Cristo Mountains, Maxwell National Wildlife Refuge lies close to the historic Santa Fe Trail. The 3,699-acre property on the Central Flyway includes reservoirs leased from the Vermejo Conservancy District that provide water for local ranchers and farmers, as well as resting and breeding habitat for seasonal waterfowl. Surrounding the lakes are wetlands where grebes and waders congregate. The plains-mesa grasslands host a variety of species not as easily seen in other locations. There are large stands of cottonwoods and elms, both in woodlots and surrounding farm structures, which support raptors and riparian species. The refuge's agricultural areas are planted to provide food for wintering and migrating birds as well as nesting species that prefer this type of habitat. The refuge has been designated an Important Bird Area by Audubon New Mexico.

Begin your birding as you drive west on NM 505, inspecting fence lines and power poles. There is a prairie dog colony on the north side of NM 505 shortly before you reach the refuge that hosts Burrowing Owl during the spring and summer months. There is a small path along the west side of A3 Road that overlooks the black-tailed prairie dog colony. Interpretive signs that line the trail provide educational material to visitors about species most often found in shortgrass prairie habitats. In the center of the trail is an observation deck with two scopes for visitors to view prairie dogs or watch the Burrowing Owls.

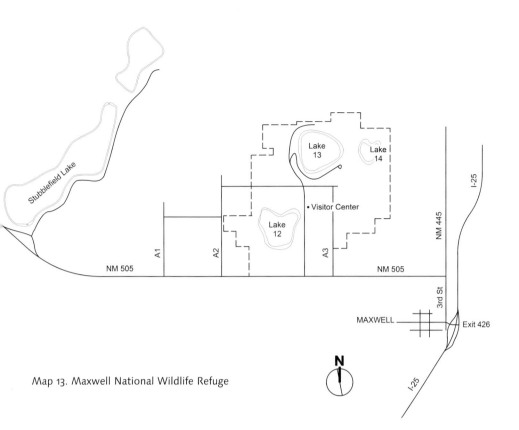

Map 13. Maxwell National Wildlife Refuge

Travel north at the refuge sign, driving slowly as you proceed toward the Visitor Center, examining the fence lines and grassy habitat on either side of the road. When the Visitor Center is not open (open 9:00 a.m. to 3:00 p.m. year-round, dependent on staff availability), pick up a map, a bird list, and other information from the nearby kiosk. Examine the large stand of cottonwood trees around the administrative buildings. Continue driving north to the parking area for Lake 13. Before following the road around the lake, take time to bird the periphery of the parking area. The primitive camping area has locations where you can access the marshes and seasonal mudflats. The road continues along the top of the levee around the western and southern edges of the lake. An arroyo adjacent to the southern levee provides habitat for songbirds, including American Tree Sparrow during the winter. A scope is necessary to adequately bird the lake.

A trail through a small cottonwood gallery on Laguna Madre Road just north of Lake 13 provides an opportunity to observe Sandhill Cranes and geese land on Lake 13 or to watch them graze in the agricultural fields

Birders at Lake 13

surrounding the trail. This trail is the entrance to the new interior auto tour route that is open seasonally on weekends in October.

After exploring the Lake 13 area, continue north a short distance to investigate the woodlots near the edge of the refuge for owls and Wild Turkey. Several roads crisscross the refuge and provide possible opportunities to observe breeding grassland species: Cassin's, Vesper, Lark, and Grasshopper Sparrows. The refuge auto tour route continues along Laguna Madre and Highline Roads. Stop at the turnouts along the way to enjoy trails through cottonwood gallery, shortgrass prairie, or black-tailed prairie dog trails. A 0.25-mile trail on Highline Road at the northeast corner of the refuge leads into the shortgrass prairie overlooking Lake 14, just east of Lake 13. Frequent sightings have included Long-billed Curlews, Swainson's Hawks, and Western Meadowlarks, as well as coyote and pronghorn. These areas are reliable locations to observe Sandhill Cranes in the fall, neotropical migrants in the spring, or raptors in the summer.

Return to NM 505 and drive west. The road makes a turn to the right shortly after crossing a marshy area. Turn right on the first dirt road, which leads to Stubblefield Lake (reservoir), owned by the Vermejo Conservancy

District. The road heads onto a dike that runs along the east side of the lake. Please respect this private property and limit your birding to the top of the levee. In recent years the water level has been quite low and at times has had no water at all.

County: Colfax

eBird Hotspots: Maxwell NWR and Stubblefield Lake

Target Birds

Common Goldeneye Large numbers spend the winter on Maxwell's Lake 13 and on Stubblefield Lake. The largest concentrations can be found between late November and the end of February.

Eared Grebe While it can be found from March through November, the highest prevalence is in May and during fall migration. It has nested in the marshy areas along Lake 13 and at Stubblefield Lake in the late summer.

Bald Eagle A number of Bald Eagles can be found between the end of October and March, with the highest concentrations in February as migrating eagles gather. Occasionally one is seen between April and September.

Rough-legged Hawk This is a fairly reliable location for Rough-legged Hawk November through January.

Long-billed Curlew In addition to those that stop over during migration, a few nest within the refuge and near Stubblefield Lake. June is a good time to look for them.

Common Goldeneye

Eastern Kingbird Maxwell is one of the reliable nesting locations for Eastern Kingbird. It is present from May through August.

Loggerhead and Northern Shrikes Both species of shrike can be seen in the area. While Loggerhead is more prevalent and generally present year-round, Northern Shrike is an irregular winter visitor between November and March.

American Tree Sparrow It is a regular winter visitor between November and February.

Lark Bunting While a few are present during the summer, it is seen regularly during spring and fall migration (NM 445, east of the refuge).

Grasshopper Sparrow Although the numbers are low, it is a regular breeder in the grasslands around the refuge and is present between May and August (at the turnoff to Lake 13).

Other Birds

Although the surface of the lakes may be frozen at times during the winter (see "Special Considerations and Hazards"), when the water is open, Maxwell NWR and Stubblefield Lake attract a wide range of wintering waterfowl. Larger concentrations of each species occur during migration. The most productive location to observe wintering waterfowl is Lake 13. Species include Snow and Cackling Geese; American Wigeon (September, October); Northern Shoveler (March and April, September and October); Northern Pintail (February and September); Green-winged Teal (September); Canvasback (early March and October); Redhead (early March); Bufflehead (April and end of November); and Hooded, Common, and Red-breasted Mergansers (January and late November).

Maxwell NWR is an excellent location for observing Ferruginous Hawk and Golden Eagle during winter months. Other regular winter species include Downy and Hairy Woodpeckers, Northern Flicker, Brown Creeper, Marsh Wren, American Pipit, and Pine Siskin.

The following are possible, but not guaranteed, during the winter: Greater White-fronted and Ross's Geese; Ring-necked Duck; Barrow's Goldeneye; Common Loon; Sharp-shinned Hawk; Merlin; and Lapland, Chestnut-collared, and McCown's Longspurs.

Waterbirds that can be seen during migration include Blue-winged and Cinnamon Teal (September); Lesser Scaup (March, April, and November–early December); Ruddy Duck; Western (September), Clark's (September-

November), and Horned Grebes (not every year); American White Pelican; and White-faced Ibis. Migrating raptors include Osprey and Cooper's Hawk.

The mudflats on all of the lakes attract a wide variety of migrating shorebirds during both spring and fall: Black-necked Stilt; Solitary Sandpiper; Greater and Lesser Yellowlegs; Willet (not every year); Marbled Godwit; Sanderling (not every year); Western, Least, Baird's, and Stilt Sandpipers; Long-billed Dowitcher; Wilson's Snipe; and Wilson's Phalarope. Pectoral Sandpiper is seen only during fall migration. Bonaparte's, Franklin's, Ring-billed, and Herring Gulls are present during migration, as is Forster's Tern.

Other migrants include White-throated Swift, Rufous Hummingbird (summer and fall), and Red-naped Sapsucker. Passerine migrants include Olive-sided, Willow, Gray, and Dusky Flycatchers; Plumbeous Vireo; Tree, Violet-green, and Bank Swallows; House Wren; Ruby-crowned Kinglet; Western Bluebird; Orange-crowned, Virginia's, MacGillivray's, Yellow-rumped (both Audubon's and Myrtle), and Wilson's Warblers; Brewer's, Song, Lincoln's, and White-crowned Sparrows; Dark-eyed Junco; Lazuli Bunting; and Western Tanager.

Species that have nested at the refuge include Blue-winged and Cinnamon Teal; Western Grebe (if water levels are suitable); American Bittern (occasionally); Swainson's Hawk; Killdeer; Spotted Sandpiper; American Avocet; Burrowing Owl; Common Nighthawk; Western Wood-Pewee; Say's Phoebe; Cassin's and Western Kingbirds; Cliff and Barn Swallows; American Robin; Northern Mockingbird; Common Yellowthroat; Yellow Warbler; Cassin's, Vesper, and Lark Sparrows; Blue Grosbeak; Brown-headed Cowbird; Bullock's Oriole; and Lesser Goldfinch.

Other birds that are regularly seen during the summer are Clark's Grebe, Double-crested Cormorant, Turkey Vulture, Northern Rough-winged Swallow, Rock Wren, Spotted Towhee, Dickcissel, Yellow-headed and Brewer's Blackbirds, and Common Grackle.

Birds present year-round include Canada Goose, Gadwall, Mallard, Ring-necked Pheasant, Wild Turkey, Pied-billed Grebe, Great Blue Heron, Northern Harrier, Red-tailed Hawk, American Kestrel, Prairie Falcon (occasional), American Coot, Eurasian Collared-Dove, Mourning Dove, Great Horned Owl, Black-billed Magpie (west of Stubblefield Lake), American Crow, Common Raven, Horned Lark, White-breasted Nuthatch, European Starling, Red-winged Blackbird, Western Meadowlark, and House Finch.

DIRECTIONS

From Las Vegas, New Mexico: From the intersection of I-25 and NM 104, travel north on I-25 approximately 80 miles to Exit 426 (Maxwell). Turn left onto NM 505 (Maxwell Avenue) and travel west less than a mile to 3rd Street, where NM 505 makes a right turn. Continue on NM 505 (3rd Street) 0.8 mile to follow NM 505 as it turns west. After 2.5 miles, turn right at the refuge sign and travel north approximately 1.2 miles to the Visitor Center.

From Cimarron Canyon State Park (3 miles east of Eagle Nest): At the eastern state park boundary on US 64, travel east through the town of Cimarron approximately 26 miles to NM 505. Turn right and travel approximately 9.5 miles to the refuge entrance road.

PARKING

Parking is available next to the Visitor Center and in a lot adjacent to Lake 13. If you park on the side of one of the refuge roads, pull off far enough to allow other vehicles to pass. At Stubblefield Lake, you can park on top of the levee.

FEES

None

SPECIAL CONSIDERATIONS AND HAZARDS

See chapter 2 for further safety guidelines.

- Rattlesnakes: Both prairie and western diamondback rattlesnakes live on the refuge.
- Summer thunderstorms: Lightning is possible, and many roads may become muddy and impassable.
- Mosquitoes: They are sometimes a problem following heavy rain.
- Winter weather conditions: The area is subject to severe winter weather conditions, which may cause the surface of the lakes to freeze and may make travel to and within the refuge difficult. Always check road conditions before traveling to the area during winter.

FACILITIES

- Accessibility: Most areas are level.
- Restrooms: Accessible restrooms are near Lake 13 and inside the Visitor Center (when open).

- Water: There is a drinking fountain inside the Visitor Center (when open) or can be purchased from gas station in Maxwell.
- Picnic tables: By Lake 13 and outside the Visitor Center
- Pets: Refuge rules require that dogs be on leash.

CAMPING

There is a primitive camping area near Lake 13.

GAS, FOOD, AND LODGING

A 24-hour gas station is available in Maxwell and also at the travel center on I-25 approximately 8 miles south of Maxwell. Limited food and beverages are available in a convenience store in the town of Maxwell; more complete food choices can be found at the travel center. Lodging is available in Springer and Raton on I-25.

CHAPTER 11

High-Elevation Specialty Locations

General Overview

The two sites described in this chapter offer opportunities to view bird species that normally can be found only at high elevations. One of these locations, reached by a strenuous hike, is above the tree line in alpine habitat. During the short period it is accessible, it hosts breeding White-tailed Ptarmigan, American Pipit, and White-crowned and Lincoln's Sparrows. The other site, the only easily reachable breeding location for Boreal Owl, is accessible by high-clearance vehicle; however, it is snow-free for only a few months.

Santa Barbara Ridge

Description

Those who have gone to this spot in search of the White-tailed Ptarmigan have referred to the hike as the "Ptarmigan Trek," which takes you to the area fairly near Jicarita Peak (one of New Mexico's highest peaks). Searching for these birds can be arduous: snowbanks often persist into June, and thunderstorms occur almost daily in July and August. However, with proper planning this hike can be extremely rewarding, with good birding the whole way and spectacular views of the Pecos Wilderness from the top of the ridge.

The ridge is located the near the north boundary of the Pecos Wilderness area in the Carson National Forest. From the Alamitos Trailhead (Serpent Lake Trail #19 at 10,412 feet) it is a 4.5-mile hike to Santa Barbara Ridge. The initial 1.5 miles go through spruce-fir woodlands, before the trail opens out to a montane meadow. At 3.5 miles you will reach a spur trail on your right that goes to Serpent Lake; stay on the main trail here. The hike from this point up to the ridge is along a rocky but well-maintained trail above the timberline. The last part of the trail is very steep. Along the way the trail

Santa Barbara Ridge (photo by Cole J. Wolf)

will gain almost 2,000 feet in elevation. Once you reach the top of the ridge, turn right and head for the small, unnamed peak. The area around this peak is the most reliable spot to find ptarmigan, but they could turn up anywhere along the ridge. Although not as steep as the hike up, the ridgetop is typical ptarmigan habitat: uneven rock fields and meadows. Ptarmigan are usually not very easy to find. Plan to spend an hour or more searching for them. The birds are well camouflaged and blend in with their rocky habitat. It is easy to lose sight of them from less than 10 feet away. They also tend to stop moving, crouch down when they sense danger, and almost never flush. Listen to recordings *before* the hike. Females with chicks especially tend to call when people approach closely.

County: Taos

eBird Hotspots: Jicarita Peak (Santa Barbara Ridge)

Target Birds

White-tailed Ptarmigan

A true tundra specialist, this species reaches the southern limit of its range in northern New Mexico. In recent years Santa Barbara Ridge in the vicinity of Jicarita Peak has proved to be the most reliable location in the state to find this species. Late June through September is typically the time of year when access to ptarmigan habitat is feasible.

Other Birds

Many high-elevation Rocky Mountain specialties can be encountered along the trail, including Dusky Grouse, American Three-toed Woodpecker, Williamson's Sapsucker, Hammond's Flycatcher, Gray Jay, Clark's Nutcracker, Golden-crowned Kinglet, Red Crossbill, and Pine and Evening Grosbeaks. At the base of the ridge where the forest ends, the willow thickets hold breeding Lincoln's and White-crowned Sparrows and possibly Wilson's Warblers. As the trail begins to ascend the ridge, birds become scarce, with American Pipit and the occasional Common Raven often the only common species in tundra habitats (although there is a chance of encountering bighorn sheep).

White-tailed Ptarmigan (photo by Jim Joseph)

DIRECTIONS

From Taos: At the intersection of US 64 and NM 68 one block east of the plaza, drive 3.5 miles south on NM 68 to the town of Rancho de Taos. Turn left onto NM 518 and travel approximately 29.5 miles (passing the town of Tres Ritos) to Forest Road 161. Turn right onto FR 161, traveling approximately 3 miles to its end at a turnaround. Follow the old gravel road at the end of the turnaround about 0.25 mile to the Serpent Lake Trailhead (#19) at Alamitos.

From Las Vegas, New Mexico (off I-25): At the intersection of Grand Avenue (US 85) and 7th Street (NM 518), travel northwest through the city on NM 518 approximately 44 miles (through the towns of Mora and Cleveland to the left turn onto FR 161).

PARKING

The road dead-ends at a large turnaround with ample parking. Please be considerate and leave room for horse trailers to use the turnaround.

FEES

None

SPECIAL CONSIDERATIONS AND HAZARDS

See chapter 2 for further safety guidelines.

- Nature of hike: The hike up can be very strenuous, and an early start is necessary to beat afternoon thunderstorms to the ridge. Determined hikers in good physical condition may be able to make it to the top of the ridge in a little over two hours, but this is an optimistic estimate for most birders. If you plan on birding along the trail, it can easily take four hours or more. There are few maintained trails on top of the ridge. Expect to spend at least six to eight hours away from your car. This hike is not recommended for those who have just flown in from sea level. Take two or three days to acclimatize to northern New Mexico's altitude.
- Altitude: Since this strenuous hike starts at over 10,000 feet and makes a 2,000-foot elevation gain, it is extremely important to take safety precautions to prevent the effects of altitude sickness.
- Thunderstorms: Thunderstorms can develop by early afternoon during the summer months. It is important to start this hike early in

High-Elevation Locations   ◀ 231

the morning so you will not be caught in an exposed area in event of lightning. Obtain weather reports before starting the hike and be prepared.

FACILITIES

- Accessibility: This hike is not recommended for anyone with mobility limitations.
- Restrooms: None available. The nearest restroom is at the Sipapu Ski and Summer Resort a short distance west of Tres Ritos.
- Water: No drinking water available. Bring enough for all day or purifying agents.
- Picnic tables: None available
- Pets: Allowed; follow appropriate National Forest and Wilderness regulations.

CAMPING

Camping is allowed at the trailhead and in the wilderness. Serpent Lake makes a convenient base camp if you decide to split the hike into two days.

GAS, FOOD, AND LODGING

The closest gas stations are in Peñasco and Taos. The nearest hotels are in Taos (less than 30 miles from the trailhead), but there is lodging available in Tres Ritos. Restaurants are located at Sipapu Ski and Summer Resort and in Peñasco.

Apache Creek

Description

This remote area of the Carson National Forest in the subalpine zone (10,000 feet) provides vehicle access to high-altitude species that cannot be seen as easily in other areas of the state. The habitat is a mixture of montane meadow grassland bordered by stands of spruce-fir forest. The valley along Forest Road 117 along the Colorado portion of the road passes through working cattle ranches where you may encounter truck vehicle traffic. The valleys also attract raptors, such as Red-tailed and Swainson's Hawks.

Shortly after reentering New Mexico and crossing the Carson National Forest boundary, there is a fork in the road and a stand of trees with a turnout on your left that can sometimes be productive for birding. Take the left

Apache Creek

fork and continue across the meadow to the point where the road passes
over Apache Creek. There is a service road almost immediately on your
right after the road heads back north again that can serve as a good walking
trail to search the uphill slope for Gray Jay and Clark's Nutcracker. Continue
north and up and around the hill in your vehicle. As the road passes
through the spruce-fir forest, listen for woodpeckers and check for activity
under the trees. The road also passes through many smaller meadows and
along the forest edge. Getting out of the car and walking along the road in
these areas will likely produce more birds. While you can continue along the
road for some distance, once the road reaches another open meadow, you
might encounter individuals with permits to harvest dead and downed trees
for fuel (May 1–December 31). There are a few wide areas along the road

where you can pull over and get out while you search for birds. If there have been recent rains in the area, it can be worth watching puddles in the road. Many bird species will visit these for water.

County: Rio Arriba

eBird Hotspots: Carson NF—Apache Creek

Target Birds

Boreal Owl This is one of the few accessible areas where it consistently nests, although it can be very difficult to locate. While it nests in cavities, it roosts in different locations each night. The nesting period extends roughly from April through August. It does not broadcast its haunting, staccato call after securing a mate, and recorded playback during the nesting period is very disruptive to this sensitive species. To increase your odds of locating one, visit during early September when it is safe to use playback. The owl may not respond with the staccato song. Instead, it may approach silently and watch you or may respond with a sharp, crisp call.

American Three-toed Woodpecker To locate areas where it is foraging, listen for its light uneven tapping sound and look for flaked bark at the base of Engelmann spruce. This woodpecker is easiest to spot in mid- to late July when it is feeding its young and is more active.

Gray Jay It tends to set up breeding and feeding territories. Once you have located a jay, one or two others are bound to appear nearby. While it is often gregarious around humans in other locations, this species may not be as friendly in this remote area. The forested area just beyond the stream crossing is fairly reliable.

Clark's Nutcracker It prefers forest edges, often seen alone or in groups of two or three.

Lincoln's Sparrow It forages on the ground in areas of dense vegetation.

Pine Grosbeak Once the nesting period is over, usually late July, it tends to gather in flocks.

Cassin's Finch It is a common breeder in the area and also tends to flock once the breeding period is over.

Other Birds

The following species are possible from mid-June through the end of August: Wild Turkey; Turkey Vulture; Swainson's and Red-tailed Hawks; Long-eared (irregular) and Great Horned Owls; Broad-tailed Hummingbird;

Boreal Owl (photo by Nancy E. Hetrick)

Williamson's and Red-naped Sapsuckers; Lewis's, Downy, and Hairy Woodpeckers; Northern Flicker; American Kestrel; Olive-sided Flycatcher (gone by mid-August); Western Wood-Pewee; Hammond's and Cordilleran Flycatchers; Warbling Vireo; Steller's Jay; American Crow; Common Raven; Violet-green Swallow; Mountain Chickadee; Red-breasted and White-breasted Nuthatches; Brown Creeper; House Wren; Golden-crowned and Ruby-crowned Kinglets; Mountain Bluebird; Hermit Thrush; Orange-crowned, Yellow-rumped, and Wilson's Warblers; Spotted Towhee; Chipping, Vesper, and White-crowned Sparrows; Dark-eyed Junco; Western Tanager; Black-headed Grosbeak; Brewer's Blackbird; Red Crossbill; Pine Siskin; and Evening Grosbeak.

During migration, you might spot Townsend's or Wilson's Warblers.

DIRECTIONS

From Chama, at the intersection of Terrace Avenue (NM 17) and 3rd Street (near the Cumbres and Toltec Scenic Railroad Station), travel 8 miles north on NM 17 to the Colorado boundary, where the highway becomes Colorado State Highway 17. Continue 5.8 miles north on CO 17, turning right on Forest Road 117. FR 117 begins at a fairly large turnout on the right between mile markers 5 and 6. The obscure brown sign for FR 117 is placed at the back of the turnout. After about 0.4 mile, FR 117 crosses the tracks for the Cumbres and Toltec Scenic Railroad. Continue 2.5 miles, leaving Rio Grande National Forest, entering Carson National Forest, and reentering New Mexico. In 0.2 mile, take a left fork onto FR 87 (do not take the right fork, FR 686). Continue another 0.2 mile to Apache Creek.

To reach Chama from Santa Fe, take US 84 north approximately 108 miles to the point where US 84 makes a sharp left. Drive straight onto NM 17 for 1.6 miles to Chama, about 2 hours.

PARKING

There is no designated parking, but there are a few turnouts. If parking where there is no turnout, be sure to allow room for another vehicle to pass, without damaging the vegetation along the road.

FEES

None

SPECIAL CONSIDERATIONS AND HAZARDS

See chapter 2 for further safety guidelines.

- Weather: Snow can be present until mid- to late June, although not every year. Summer thunderstorms can make the roads muddy and impassable. Take care not to get stuck.
- Road conditions: The dirt road is rutted and not well maintained. *A high-clearance vehicle is a must.*
- Remote location: This area is extremely remote. There is no cell phone coverage.
- Black bears: Bears are possible. Do not leave food unsecured.
- Cougars: Cougars are found in this area and are most active from dusk to dawn. Birders searching for owls need to be aware of the possibility of encountering a cougar.
- Hunting season: Bow hunting begins mid-September. If you are birding at that time, dress in bright orange to be seen.

- Accessibility: Walking along the road is fairly easy. Walking up the slopes to seek out a specific species might require walking over uneven terrain.
- Restrooms: The closest restrooms are in Chama.
- Water: None available
- Picnic tables: None available
- Pets: Carson National Forest requires that dogs be on leash.

CAMPING

Primitive camping is allowed in this area. One good area is near the service road just past the stream crossing. Many birders sleep in their vehicles. See the US Forest Service website (http://www.fs.usda.gov/activity/carson/recreation/camping-cabins) for guidelines.

GAS, FOOD, AND LODGING

There are abundant options for gas, food, and lodging in Chama.

American Birding Association's Code of Ethics

Everyone who enjoys birds and birding must always respect wildlife, its environment, and the rights of others. In any conflict of interest between birds and birders, the welfare of the birds and their environment comes first. The American Birding Association's Code of Birding Ethics may be freely reproduced for distribution/dissemination. Please acknowledge the role of ABA in developing and promoting this code with a link to the ABA Wwebsite using the URL http://www.aba.org. Thank you.

Code of Birding Ethics

1. Promote the welfare of birds and their environment.

1(a) Support the protection of important bird habitat.

1(b) To avoid stressing birds or exposing them to danger, exercise restraint and caution during observation, photography, sound recording, or filming.

Limit the use of recordings and other methods of attracting birds, and never use such methods in heavily birded areas or for attracting any species that is Threatened, Endangered, or of Special Concern, or is rare in your local area.

Keep well back from nests and nesting colonies, roosts, display areas, and important feeding sites. In such sensitive areas, if there is a need for extended observation, photography, filming, or recording, try to use a blind or hide and take advantage of natural cover.

Use artificial light sparingly for filming or photography, especially for close-ups.

1(c) Before advertising the presence of a rare bird, evaluate the potential for disturbance to the bird, its surroundings, and other people in the area, and proceed only if access can be controlled, disturbance minimized, and permission has been obtained from private landowners. The sites of rare nesting birds should be divulged only to the proper conservation authorities.

1(d) Stay on roads, trails, and paths where they exist; otherwise, keep habitat disturbance to a minimum.

2. Respect the law, and the rights of others.

2(a) Do not enter private property without the owner's explicit permission.

2(b) Follow all laws, rules, and regulations governing use of roads and public areas, both at home and abroad.

2(c) Practice common courtesy in contacts with other people. Your exemplary behavior will generate goodwill with birders and nonbirders alike.

3. Ensure that feeders, nest structures, and other artificial bird environments are safe.

3(a) Keep dispensers, water, and food clean and free of decay or disease. It is important to feed birds continually during harsh weather.

3(b) Maintain and clean nest structures regularly.

3(c) If you are attracting birds to an area, ensure the birds are not exposed to predation from cats and other domestic animals or dangers posed by artificial hazards.

4. Group birding, whether organized or impromptu, requires special care.

Each individual in the group, in addition to the obligations spelled out in Items #1 and #2, has responsibilities as a Group Member.

4(a) Respect the interests, rights, and skills of fellow birders, as well as people participating in other legitimate outdoor activities. Freely share your knowledge and experience, except where code 1(c) applies. Be especially helpful to beginning birders.

4(b) If you witness unethical birding behavior, assess the situation and intervene if you think it prudent. When interceding, inform the person(s) of the inappropriate action and attempt, within reason, to have it stopped. If the behavior continues, document it, and notify appropriate individuals or organizations.

Group Leader Responsibilities (amateur and professional trips and tours).

4(c) Be an exemplary ethical role model for the group. Teach through word and example.

4(d) Keep groups to a size that limits impact on the environment and does not interfere with others using the same area.

4(e) Ensure everyone in the group knows of and practices this code.

4(f) Learn and inform the group of any special circumstances applicable to the areas being visited (e.g., no tape recorders allowed).

4(g) Acknowledge that professional tour companies bear a special responsibility to place the welfare of birds and the benefits of public knowledge ahead of the company's commercial interests. Ideally, leaders should keep track of tour sightings, document unusual occurrences, and submit records to appropriate organizations.

Please follow this Code and distribute and teach it to others.

Annotated Checklist

Most of the 276 species listed here are those mentioned in one or more of the site descriptions. Also included are species that are seen regularly in north-central New Mexico, but not with any regularity at any of the sites. Species documented in north-central New Mexico, but considered rarities or vagrants, are not included. Report any species not included on the following list to the New Mexico Bird Records Committee (see "Local Birding Information and Resources" in chapter 2).

The following abbreviations are used in this checklist:

- KKTRNM: Kasha-Katuwe Tent Rocks National Monument
- LVNWR: Las Vegas National Wildlife Refuge
- NF: National Forest
- NM: National Monument
- NWR: National Wildlife Refuge
- OOFL: Ohkay Owingeh Fishing Lakes
- RDAC: Randall Davey Audubon Center
- SP: State Park
- VCNP: Valles Caldera National Preserve

The sequence and names conform to the American Ornithologists' Union's *Checklist of North American Birds,* 7th Edition (1998) as amended through its 54th Supplement (September 2013). A species marked with an asterisk (*) indicates a species designated by the New Mexico Avian Partners as the highest concern. These abundance terms are used in the checklist:

- Abundant: Referring to a species that is numerous
- Common: Likely to be seen or heard in suitable habitat
- Uncommon: Present, but not certain to be seen
- Occasional: Seen only a few times during a season
- Casual: Not seen every year

- ☐ **Greater White-fronted Goose** (*Anser albifrons*): Occasional during winter at Maxwell NWR and LVNWR.
- ☐ **Snow Goose** (*Chen caerulescens*): A common migrant and abundant during winter at LVNWR. Arrives in mid-October and can be seen in abundance through mid-March, although stragglers may remain as late as early April.
- ☐ **Ross's Goose** (*Chen rossii*): An uncommon migrant and locally abundant mingling with the flocks of Snow Geese at LVNWR between November and mid-March.
- ☐ **Cackling Goose** (*Branta hutchinsii*): Uncommon during winter at Maxwell NWR and LVNWR.
- ☐ **Canada Goose** (*Branta canadensis*): Common year-round resident in north-central New Mexico. By mid-October, becomes abundant as wintering flocks arrive, remaining until mid-March. Can be seen foraging in agricultural fields, on the rivers, or at any of the lakes.
- ☐ **Tundra Swan** (*Cygnus columbianus*): Seen occasionally between November and mid-March, primarily at LVNWR and Maxwell NWR.
- ☐ **Wood Duck** (*Aix sponsa*): Found occasionally in areas with small lakes and good riparian habitat. Look for it along the Rio Grande at OOFL.
- ☐ **Gadwall** (*Anas strepera*): Uncommon winter visitor to Cochiti Lake and occasional migrant at Maxwell NWR and LVNWR.
- ☐ **American Wigeon** (*Anas americana*): Uncommon winter visitor to Cochiti Lake and occasional spring migrant at LVNWR.
- ☐ **Mallard** (*Anas platyrhynchos*): Abundant year-round almost anywhere there is water.
- ☐ **Blue-winged Teal** (*Anas discors*): Uncommon migrant and summer resident and breeder between April and October at LVNWR, Storrie Lake SP, Maxwell NWR, and Cochiti Lake.
- ☐ **Cinnamon Teal** (*Anas cyanoptera*): Occasional in spring at Cochiti Lake; summer and fall at LVNWR.
- ☐ **Northern Shoveler** (*Anas clypeata*): Common in winter and occasional year-round on lakes in north-central New Mexico.
- ☐ **Northern Pintail** (*Anas acuta*): Common in winter at Cochiti Lake, Maxwell NWR, and LVNWR.
- ☐ **Green-winged Teal** (*Anas crecca*): Uncommon in winter at LVNWR and Cochiti Lake.

☐ Canvasback (*Aythya valisineria*): Occasional in winter at LVNWR and Cochiti Lake.

☐ Redhead (*Aythya americana*): Occasional in fall and winter at Maxwell NWR, LVNWR, and Cochiti Lake.

☐ Ring-necked Duck (*Aythya collaris*): Uncommon winter visitor at Cochiti Lake and LVNWR.

☐ Greater Scaup (*Aythya marila*): Casual in winter at LVNWR.

☐ Lesser Scaup (*Aythya affinis*): Uncommon in winter at Cochiti Lake, Maxwell NWR, and LVNWR.

☐ Bufflehead (*Bucephala albeola*): A target bird at OOFL. Common in winter at Cochiti Lake, Maxwell NWR, and LVNWR.

☐ Common Goldeneye (*Bucephala clangula*): A target bird at OOFL. Uncommon in winter at Cochiti Lake and Maxwell NWR.

☐ Barrow's Goldeneye (*Bucephala islandica*): Casual between November and mid-March at Maxwell NWR, along the Rio Grande between Taos and Española.

☐ Hooded Merganser (*Lophodytes cucullatus*): Occasional in winter at Cochiti Lake and LVNWR.

☐ Common Merganser (*Mergus merganser*): Uncommon between November and mid-March at OOFL, Cochiti Lake, Maxwell NWR, and LVNWR.

☐ Red-breasted Merganser (*Mergus serrator*): Casual winter visitor at Maxwell NWR and Cochiti Lake; casual migrant at other locations.

☐ Ruddy Duck (*Oxyura jamaicensis*): Uncommon in winter at Eagle Nest Lake SP, LVNWR, Maxwell NWR, and Cochiti Lake.

☐ Scaled Quail (*Callipepla squamata*): Occasional in desert scrub and piñon-juniper habitat at Cochiti Lake and White Rock.

☐ Ring-necked Pheasant (*Phasianus colchicus*): Occasional year-round at OOFL, Maxwell NWR, and La Cieneguilla Marsh.

☐ White-tailed Ptarmigan (*Lagopus leucura*): Occasional year-round above the tree line, including Santa Barbara Ridge and along the Bull-of-the-Woods Trail above Taos Ski Valley when snow is not present.

☐ Dusky Grouse (*Dendragapus obscurus*): Occasional between April and October at high elevations, including Santa Fe Ski Area, Pajarito Mountain, and Taos Ski Valley.

☐ Wild Turkey (*Meleagris gallopavo*): Occasional to locally common year-round. Reliable at Cimarron Canyon SP.

- ☐ Pacific Loon (*Gavia pacifica*): Casual in fall at a variety of lakes.
- ☐ Common Loon (*Gavia immer*): Occasional at lakes and reservoirs, primarily October to February.
- ☐ Pied-billed Grebe (*Podilymbus podiceps*): Common in winter and uncommon at other times on lakes and reservoirs.
- ☐ Horned Grebe (*Podiceps auritus*): Casual, primarily in November at lakes and reservoirs.
- ☐ Eared Grebe (*Podiceps nigricollis*): Common migrant on selected lakes and reservoirs from April through June as well as October and November. Breeding does not occur at sites in this book.
- ☐ Western Grebe (*Aechmophorus occidentalis*): Uncommon to locally abundant migrant during April and mid-October through November at Cochiti Lake, Heron Lake SP, Eagle Nest Lake SP, and lakes along I-25. Uncommon during June and July. Breeding often late after summer rains raise water in lakes to sufficient level.
- ☐ Clark's Grebe (*Aechmophorus clarkii*): Uncommon to locally abundant migrant in April and mid-October through November at Cochiti Lake, Heron Lake SP, Eagle Nest Lake SP, and lakes along I-25. Uncommon during June and July. Breeding often late after summer rains raise water in lakes to sufficient level.
- ☐ Double-crested Cormorant (*Phalacrocorax auritus*): Uncommon to common migrant with highest abundance from October through November and locally common at lakes/reservoirs.
- ☐ American White Pelican (*Pelecanus erythrorhynchos*): Uncommon to abundant migrant between July and October. The largest concentration can be found at Eagle Nest Lake SP, where it rests following breeding and before migrating farther south.
- ☐ Great Blue Heron (*Ardea herodias*): Uncommon to common May through November in wetlands, on edges of lakes, and along rivers throughout north-central New Mexico.
- ☐ Snowy Egret (*Egretta thula*): Occasional migrant at lakes and wetlands.
- ☐ Cattle Egret (*Bubulcus ibis*): Casual during summer at lakes and wetlands.
- ☐ Green Heron (*Butorides virescens*): Occasional in summer at lakes and wetlands along the Rio Grande.

- [ ] Black-crowned Night-Heron (*Nycticorax nycticorax*): Occasional along creeks and rivers May through August.
- [ ] White-faced Ibis (*Plegadis chihi*): Uncommon migrant mid-April through mid-May and mid-August through mid-September at lakes, including Heron Lake SP, LVNWR, and Maxwell NWR.
- [ ] Turkey Vulture (*Cathartes aura*): Common migrant and summer resident in all areas of north-central New Mexico.
- [ ] Osprey (*Pandion haliaetus*): Uncommon migrant and locally common April through September at Cochiti Lake, El Vado Lake SP, and Heron Lake SP.
- [ ] Bald Eagle (*Haliaeetus leucocephalus*): Uncommon and locally common winter resident at lakes and along rivers throughout north-central New Mexico. Nests at locations in Rio Arriba and Colfax Counties.
- [ ] Northern Harrier (*Circus cyaneus*): Uncommon migrant and winter visitor mid-September through mid-March over wetlands and agricultural areas.
- [ ] Sharp-shinned Hawk (*Accipiter striatus*): Uncommon migrant and winter visitor, primarily in Sangre de Cristo and Jemez Mountains and along Rio Grande.
- [ ] Cooper's Hawk (*Accipiter cooperii*): Uncommon in lower elevations in Sangre de Cristo and Jemez Mountains and in riparian areas along major rivers.
- [ ] Northern Goshawk (*Accipiter gentilis*): Occasional year-round resident of the Jemez and Sangre de Cristo Mountains.
- [ ] Swainson's Hawk (*Buteo swainsoni*): Uncommon migrant and summer resident in grasslands and agricultural areas.
- [ ] Zone-tailed Hawk (*Buteo albonotatus*): Occasional in the Jemez Mountains near Los Alamos, where one or two pairs nest each year.
- [ ] Red-tailed Hawk (*Buteo jamaicensis*): Widespread throughout north-central New Mexico.
- [ ] Ferruginous Hawk* (*Buteo regalis*): Uncommon migrant and winter resident in grassland locations.
- [ ] Rough-legged Hawk (*Buteo lagopus*): Uncommon winter visitor at Maxwell NWR and LVNWR.
- [ ] Golden Eagle (*Aquila chrysaetos*): Occasional and locally uncommon year-round, including Rio Grande del Norte NM and Angel Fire.

- ☐ **Virginia Rail** (*Rallus limicola*): Occasional year-round at cattail marshes, including Fred Baca Park, La Cieneguilla Marsh, and RDAC.
- ☐ **Sora** (*Porzana carolina*): Casual fall migrant at wetlands in north-central New Mexico.
- ☐ **American Coot** (*Fulica americana*): Common in low numbers on lakes during summer and winter. Abundant migrant in October and November.
- ☐ **Sandhill Crane** (*Grus canadensis*): Common to abundant migrant through north-central New Mexico and common during winter at LVNWR. Begins to arrive in mid-October and has departed by the end of March.
- ☐ **Black-necked Stilt** (*Himantopus mexicanus*): Casual during spring migration, primarily along I-25 lakes.
- ☐ **American Avocet** (*Recurvirostra americana*): Uncommon migrant and locally common summer resident. Breeds some years at Maxwell NWR.
- ☐ **Semipalmated Plover** (*Charadrius semipalmatus*): Occasional migrant at Springer Lake or Maxwell NWR.
- ☐ **Killdeer** (*Charadrius vociferus*): Common migrant and summer resident in fields and at edges of lakes.
- ☐ **Mountain Plover*** (*Charadrius montanus*): Not at any sites in this book, but locally casual in Taos County northeast of Tres Piedras and common on the Vermejo Ranch (private) near Cimarron.
- ☐ **Spotted Sandpiper** (*Actitis macularius*): Uncommon and locally common migrant and breeder along rivers and at lakes.
- ☐ **Solitary Sandpiper** (*Tringa solitaria*): Seen occasionally during fall migration, primarily at Maxwell NWR.
- ☐ **Greater Yellowlegs** (*Tringa melanoleuca*): Occasional migrant through north-central New Mexico in both spring and fall, primarily at lakes along I-25.
- ☐ **Willet** (*Tringa semipalmata*): Occasional migrant through north-central New Mexico mid-April through mid-May, primarily at lakes along I-25.
- ☐ **Lesser Yellowlegs** (*Tringa flavipes*): Uncommon migrant through north-central New Mexico in both spring and fall, primarily at lakes along I-25.

- [ ] Upland Sandpiper (*Bartramia longicauda*): Casual migrant in the eastern plains that has been reported at Maxwell NWR and LVNWR.
- [ ] Long-billed Curlew* (*Numenius americanus*): Uncommon migrant, primarily at lakes along I-25, and a locally common breeder near Stubblefield Lake and Maxwell NWR.
- [ ] Marbled Godwit (*Limosa fedoa*): Occasional migrant at lakes along I-25.
- [ ] Stilt Sandpiper (*Calidris himantopus*): Occasional migrant mid-July through September at Springer Lake and Maxwell NWR.
- [ ] Sanderling (*Calidris alba*): Occasional migrant in May and September at Springer Lake and Maxwell NWR.
- [ ] Baird's Sandpiper (*Calidris bairdii*): Uncommon migrant, primarily mid-July through early October, at lakes along I-25.
- [ ] Least Sandpiper (*Calidris minutilla*): Uncommon migrant March through mid-May and mid-July through September at Cochiti Lake and along I-25.
- [ ] White-rumped Sandpiper (*Calidris fuscicollis*): Casual late-spring migrant along the eastern plains. Reported at Springer Lake and Maxwell NWR.
- [ ] Pectoral Sandpiper (*Calidris melanotos*): Occasional migrant mid-August through early October at sites along I-25.
- [ ] Semipalmated Sandpiper (*Calidris pusilla*): Casual migrant at lakes along I-25.
- [ ] Western Sandpiper (*Calidris mauri*): Occasional migrant April through mid-May and mid-July through September at lakes along I-25.
- [ ] Long-billed Dowitcher (*Limnodromus scolopaceus*): Uncommon migrant in both spring and fall at Cochiti Lake and lakes along I-25.
- [ ] Wilson's Snipe (*Gallinago delicata*): Uncommon migrant and winter resident in lowland areas. Evidence of breeding activity in wet meadows south of Chama.
- [ ] Wilson's Phalarope (*Phalaropus tricolor*): Uncommon spring and fall migrant that can be seen April through May and mid-July through mid-September at Cochiti Lake and along I-25.
- [ ] Red-necked Phalarope (*Phalaropus lobatus*): Casual fall migrant through the eastern plains at Maxwell NWR and LVNWR.
- [ ] Sabine's Gull (*Xema sabini*): Casual fall migrant through the eastern plains at Maxwell NWR and LVNWR.

- [ ] **Bonaparte's Gull** (*Chroicocephalus philadelphia*): Casual migrant in early spring and late fall at Maxwell NWR, LVNWR, and Cochiti Lake.
- [ ] **Franklin's Gull** (*Leucophaeus pipixcan*): Uncommon spring migrant at lakes, including Cochiti Lake, Heron Lake, Springer Lake, and Maxwell NWR.
- [ ] **Ring-billed Gull** (*Larus delawarensis*): Common migrant and winter visitor at lakes. Casual in summer.
- [ ] **California Gull** (*Larus californicus*): Occasional at a variety of lakes, including Eagle Nest SP, Cochiti Lake, and Heron Lake, where a pair nested for the first time in 2013.
- [ ] **Herring Gull** (*Larus argentatus*): Casual at lakes in winter.
- [ ] **Black Tern** (*Chlidonias niger*): Casual fall migrant at lakes along I-25, including Maxwell NWR, Springer Lake, and LVNWR.
- [ ] **Common Tern** (*Sterna hirundo*): Casual fall migrant at Stubblefield Lake.
- [ ] **Forster's Tern** (*Sterna forsteri*): Casual spring and fall migrant seen at locations including Heron Lake SP, OOFL, and Springer Lake.
- [ ] **Rock Pigeon** (*Columba livia*): Abundant in urban areas.
- [ ] **Band-tailed Pigeon** (*Patagioenas fasciata*): Uncommon summer resident in the Sangre de Cristo Mountains. Locally common in Los Alamos along the Perimeter Trail in Upper Rendija Canyon.
- [ ] **Eurasian Collared-Dove** (*Streptopelia decaocto):* Common introduced species, now well established and widespread throughout north-central New Mexico.
- [ ] **White-winged Dove** (*Zenaida asiatica*): Common to abundant north to Española.
- [ ] **Mourning Dove** (*Zenaida macroura*): Common migrant and summer resident May through September in grasslands, agricultural areas, and urban centers.
- [ ] **Yellow-billed Cuckoo** (*Coccyzus americanus*): Casual migrant along the upper Rio Grande.
- [ ] **Greater Roadrunner** (*Geococcyx californianus*): Uncommon year-round, primarily in areas with upland desert scrub habitat.
- [ ] **Barn Owl** (*Tyto alba*): Occasional April through September at LVNWR and Maxwell NWR.
- [ ] **Flammulated Owl*** (*Psiloscops flammeolus*): Uncommon summer resident in Los Alamos and Hyde Memorial SP.

- [ ] **Western Screech-Owl** (*Megascops kennicottii*): Uncommon year-round resident in riparian habitats.
- [ ] **Great Horned Owl** (*Bubo virginianus*): Uncommon to common year-round in a wide range of habitats. Easiest to see during breeding period.
- [ ] **Northern Pygmy-Owl** (*Glaucidium gnoma*): Occasional year-round resident recorded most often at RDAC and near Bandelier NM Visitor Center.
- [ ] **Burrowing Owl** (*Athene cunicularia*): Uncommon summer resident and breeder in grasslands. The most consistent location is Maxwell NWR.
- [ ] **Long-eared Owl** (*Asio otus*): Casual species reported most often at Maxwell NWR.
- [ ] **Boreal Owl** (*Aegolius funereus*): Occasional year-round resident in a few remote high-elevation locations. The most accessible spot is Apache Creek in Carson NF.
- [ ] **Northern Saw-whet Owl** (*Aegolius acadicus*): Uncommon high-elevation resident. Pajarito Mountain and Holy Ghost Campground are good locations.
- [ ] **Common Nighthawk** (*Chordeiles minor*): Uncommon migrant and summer resident in grasslands, meadows, and agricultural areas from May through September.
- [ ] **Common Poorwill** (*Phalaenoptilus nuttallii*): Uncommon summer resident and breeder in foothills, canyons, and mountains; more easily heard than seen from May through September.
- [ ] **Black Swift** (*Cypseloides niger*): Uncommon migrant and summer breeder at Jemez Falls from late June through September. Has been reported at Nambe Falls on the Nambe Pueblo north of Santa Fe.
- [ ] **White-throated Swift** (*Aeronautes saxatalis*): Common in summer near cliffs and canyons, including Los Alamos, KKTRNM, and Rio Grande del Norte NM.
- [ ] **Black-chinned Hummingbird** (*Archilochus alexandri*): Common summer resident and breeder at low to midelevations between April and September.
- [ ] **Broad-tailed Hummingbird** (*Selasphorus platycercus*): Common summer resident and breeder above 6,000 feet.

- [ ] Rufous Hummingbird (*Selasphorus rufus*): Common fall migrant starting in July and continuing through September.
- [ ] Calliope Hummingbird (*Selasphorus calliope*): Uncommon migrant mid-July through mid-September.
- [ ] Belted Kingfisher (*Megaceryle alcyon*): Uncommon year-round along rivers and ponds.
- [ ] Lewis's Woodpecker* (*Melanerpes lewis*): Uncommon year-round resident in open-canopy, riparian habitat. Chama is the best location to look for one.
- [ ] Acorn Woodpecker (*Melanerpes formicivorus*): Locally common year-round in areas of Los Alamos.
- [ ] Williamson's Sapsucker (*Sphyrapicus thyroideus*): Uncommon resident of mixed conifer forests. The largest concentration is in the Jemez Mountains.
- [ ] Red-naped Sapsucker (*Sphyrapicus nuchalis*): Uncommon summer resident and breeder in spruce-fir, mixed conifer, and ponderosa pine forests. During fall and winter, it moves to lower elevations, where it prefers elm trees.
- [ ] Ladder-backed Woodpecker (*Picoides scalaris*): Uncommon year-round resident and breeder in desert scrub habitat. Be alert for it at the Cochiti Lake area and Gallinas Canyon Trail at LVNWR.
- [ ] Downy Woodpecker (*Picoides pubescens*): Common year-round resident and breeder from riparian areas along rivers to high-elevation locations.
- [ ] Hairy Woodpecker (*Picoides villosus*): Common year-round resident and breeder at higher elevations.
- [ ] American Three-toed Woodpecker (*Picoides dorsalis*): Uncommon year-round resident and breeder above 10,000 feet, preferring burned or dying Engelmann spruce. Search for it at Pajarito Mountain, Taos Ski Valley, and Apache Creek.
- [ ] Northern (red-shafted) Flicker (*Colaptes auratus cafer*): Widespread common year-round breeder and common in winter at lower elevations.
- [ ] American Kestrel (*Falco sparverius*): Common year-round in open areas, including Cochiti Lake, Ghost Ranch, and LVNWR.
- [ ] Merlin (*Falco columbarius*): Occasional in winter in grasslands and agricultural areas, including LVNWR.

☐ Peregrine Falcon (*Falco peregrinus*): Uncommon from March through September in rocky canyons or areas with bluffs. Uncommon in winter in open areas, generally along rivers or wetlands.

☐ Prairie Falcon (*Falco mexicanus*): Uncommon in grasslands near cliffs during breeding and farther afield the rest of the year. A good location is LVNWR.

☐ Olive-sided Flycatcher (*Contopus cooperi*): Uncommon at high-elevation coniferous forests in Sangre de Cristo and Jemez Mountains between May and mid-September.

☐ Western Wood-Pewee (*Contopus sordidulus*): Common summer resident and breeder in riparian and ponderosa pine woodlands May through mid-September.

☐ Willow Flycatcher (*Empidonax traillii*): Uncommon migrant and breeder. The federally designated endangered subspecies, "Southwestern" (*E. t. extimus*), has been documented to occur in the marsh southwest of Camino del Medio along the Rio de Fernando de Taos (Fred Baca Park).

☐ Hammond's Flycatcher (*Empidonax hammondii*): Uncommon migrant and summer resident in mixed conifer woodlands. Often it is confused with Dusky Flycatcher. The primary identifier is its vocalization.

☐ Gray Flycatcher (*Empidonax wrightii*): Occasional migrant and uncommon summer resident and breeder in piñon-juniper habitat in the Los Alamos area, the Santa Fe foothills, and Gallinas Canyon near Las Vegas from April through September.

☐ Dusky Flycatcher (*Empidonax oberholseri*): Uncommon migrant and summer resident in mixed conifer and aspen woodlands between May and mid-September.

☐ Cordilleran Flycatcher (*Empidonax occidentalis*): Common summer resident and breeder in high-elevation coniferous forests. Arrives in May and remains through September throughout the Sangre de Cristo and Jemez Mountains.

☐ Black Phoebe (*Sayornis nigricans*): Common year-round resident and breeder throughout the north-central New Mexico riparian lowlands.

☐ Eastern Phoebe (*Sayornis phoebe*): Locally common at Villanueva SP between May and July. Casual elsewhere.

- ☐ **Say's Phoebe** (*Sayornis saya*): Common in grasslands and agricultural areas between March/April and October.
- ☐ **Ash-throated Flycatcher** (*Myiarchus cinerascens*): Common summer resident and breeder at midlevel locations from late April through mid-August.
- ☐ **Cassin's Kingbird** (*Tyrannus vociferans*): Uncommon summer resident and breeder April through September, primarily in desert scrub, piñon-juniper, and grassland habitats at higher elevations than the Western Kingbird, although their ranges overlap at some locations.
- ☐ **Western Kingbird** (*Tyrannus verticalis*): Common summer resident and breeder primarily in grassland and agricultural habitat mid-April through September.
- ☐ **Eastern Kingbird** (*Tyrannus tyrannus*): Common summer resident at Maxwell NWR between mid-May and August and casual migrant at OOFL.
- ☐ **Loggerhead Shrike** (*Lanius ludovicianus*): Uncommon year-round resident and breeder in grass and farmlands.
- ☐ **Northern Shrike** (*Lanius excubitor*): Uncommon winter resident, primarily at LVNWR and Maxwell NWR.
- ☐ **Gray Vireo*** (*Vireo vicinior*): Occasional summer resident and breeder from mid- to late April through mid-August. Most reliable nesting locations in a combined piñon-juniper and desert scrub habitat on the Caja del Rio Plateau and Diablo Canyon west of Santa Fe. The best time to spot one is when the male is singing on territory. Once it starts feeding the fledglings, it stops singing.
- ☐ **Plumbeous Vireo** (*Vireo plumbeus*): Migrant and a common summer resident in mountain canyons with deciduous understory between late April and mid-September.
- ☐ **Cassin's Vireo** (*Vireo cassinii*): Occasional fall (primarily September) migrant at RDAC and in Los Alamos.
- ☐ **Warbling Vireo** (*Vireo gilvus*): Migrant and a common summer resident and breeder in high-elevation forests in the Jemez and Sangre de Cristo Mountains between early May and the end of September. Overlaps with Plumbeous Vireo at lower elevations.
- ☐ **Gray Jay** (*Perisoreus canadensis*): Uncommon in high-altitude mixed conifer and spruce-fir habitats, including the Sandia Ski Area and Taos Ski Valley.

Northern Shrike (photo by Nancy E. Hetrick)

- [ ] Pinyon Jay* (*Gymnorhinus cyanocephalus*): Roving widely in flocks, it is an uncommon year-round resident and breeder in piñon-juniper habitat.
- [ ] Steller's Jay (*Cyanocitta stelleri*): Common year-round resident and breeder in the Jemez and Sangre de Cristo Mountains.
- [ ] Blue Jay (*Cyanocitta cristata*): Casual wanderer into north-central New Mexico.
- [ ] Western (Woodhouse's) Scrub-Jay (*Aphelocoma californica woodhouseii*): Common year-round resident in piñon-juniper habitat.
- [ ] Clark's Nutcracker (*Nucifraga columbiana*): Uncommon year-round resident and breeder primarily at altitudes above 8,000 feet, where its presence is variable.
- [ ] Black-billed Magpie (*Pica hudsonia*): Common year-round resident in riparian and populated areas throughout north-central New Mexico along the major rivers.
- [ ] American Crow (*Corvus brachyrhynchos*): Abundant year-round resident and breeder that feeds in large flocks and in communal roosts in winter.

Black-billed Magpie

- [ ] Chihuahuan Raven (*Corvus cryptoleucus*): Occasional in September at LVNWR, identified primarily by its call.
- [ ] Common Raven (*Corvus corax*): Abundant year-round resident and breeder in a wide range of habitats.
- [ ] Horned Lark (*Eremophila alpestris*): Common migrant; summer and winter resident in disturbed grasslands.
- [ ] Purple Martin (*Progne subis*): Casual migrant, primarily west of the Sangre de Cristo Mountains.
- [ ] Tree Swallow (*Tachycineta bicolor*): Common breeder at Chama, Eagle Nest SP, and upper Pecos River and uncommon migrant between April and mid-September.
- [ ] Violet-green Swallow (*Tachycineta thalassina*): Migrant and locally common summer resident and breeder at high elevations.
- [ ] Northern Rough-winged Swallow (*Stelgidopteryx serripennis*): Common migrant and uncommon summer resident at rivers and lakes between April and mid-September.
- [ ] Bank Swallow (*Riparia riparia*): Occasional migrant and breeder along the Rio Grande.
- [ ] Cliff Swallow (*Petrochelidon pyrrhonota*): Common migrant and summer resident, nesting on cliffs and under bridges between March and the end of August.
- [ ] Barn Swallow (*Hirundo rustica*): Abundant migrant, summer resident, and breeder between mid-March and early October. In addition to gleaning insects over ponds and wetlands, it patrols fields, the foothills, and mountain meadows.
- [ ] Black-capped Chickadee (*Poecile atricapillus*): Common year-round resident in cottonwood bosque. Possible at any site along the Rio Grande.
- [ ] Mountain Chickadee (*Poecile gambeli*): Common year-round resident and breeder in coniferous forests; however, uncommon at lower elevations during the winter.
- [ ] Juniper Titmouse* (*Baeolophus ridgwayi*): Common year-round resident of piñon-juniper and pine-oak habitats.
- [ ] Bushtit (*Psaltriparus minimus*): Common and widespread year-round in piñon-juniper and shrubby habitats.
- [ ] Red-breasted Nuthatch (*Sitta canadensis*): Uncommon year-round

resident and breeder in the Sangre de Cristo Mountains, although occasionally wanders through the lowlands during winter.

- ☐ **White-breasted Nuthatch** (*Sitta carolinensis*): Common year-round resident and breeder at a wide range of elevations.
- ☐ **Pygmy Nuthatch** (*Sitta pygmaea*): Uncommon year-round resident and breeder in ponderosa pine woodlands.
- ☐ **Brown Creeper** (*Certhia americana*): Uncommon in coniferous forests in the Sangre de Cristo and Jemez Mountains. Occasional at lower elevations in winter.
- ☐ **Rock Wren** (*Salpinctes obsoletus*): Common in same locations as Canyon Wren between March and October.
- ☐ **Canyon Wren** (*Catherpes mexicanus*): Uncommon to common in rocky canyons year-round, including KKTRNM, Rio Grande del Norte NM, and Los Alamos.
- ☐ **House Wren** (*Troglodytes aedon*): Common in mixed conifer and spruce-fir shrubby areas between late April and late September.
- ☐ **Winter Wren** (*Troglodytes hiemalis*): Casual in winter in dense riparian habitat, including RDAC, Fred Baca Park, and Monastery Lake.
- ☐ **Marsh Wren** (*Cistothorus palustris*): Uncommon migrant and winter resident in cattail marshes, including Fred Baca Park, RDAC, and OOFL.
- ☐ **Bewick's Wren** (*Thryomanes bewickii*): Common year-round resident and breeder in the understory in riparian and piñon-juniper woodlands.
- ☐ **Blue-gray Gnatcatcher** (*Polioptila caerulea*): Uncommon summer resident and migrant in piñon-juniper canyons, including Los Alamos and Ghost Ranch between April and September.
- ☐ **American Dipper** (*Cinclus mexicanus*): Uncommon along all fast-moving rivers year-round.
- ☐ **Golden-crowned Kinglet** (*Regulus satrapa*): Occasional in mixed conifer and spruce-fir habitat year-round. A few move to lower altitudes in winter.
- ☐ **Ruby-crowned Kinglet** (*Regulus calendula*): Common summer resident in mixed conifer habitat. Migrates to lower elevations in the fall and spends the winter at sites along the Rio Grande and Pecos River.
- ☐ **Eastern Bluebird** (*Sialia sialis*): Casual in winter at OOFL and Bandelier NM.

- [ ] **Western Bluebird** (*Sialia mexicana*): Uncommon nester in ponderosa pine habitat and in winter along the Rio Grande, Rio Chama, and RDAC.
- [ ] **Mountain Bluebird** (*Sialia currucoides*): Uncommon at edge of ponderosa pine woodlands during breeding and in montane grasslands for the remainder of the year.
- [ ] **Townsend's Solitaire** (*Myadestes townsendi*): Occasional breeder in spruce-fir and mixed conifer habitats, including Taos Ski Valley. Uncommon during winter in piñon-juniper habitat.
- [ ] **Veery** (*Catharus fuscescens*): Previously a reliable species in Chama, it has been casual for a number of years.
- [ ] **Swainson's Thrush** (*Catharus ustulatus*): Casual in Chama, along the Rio Pueblo and Rio Hondo near Taos.
- [ ] **Hermit Thrush** (*Catharus guttatus*): Uncommon breeder in mixed conifer and spruce-fir habitats. Migrates to lower elevations in winter, including RDAC.
- [ ] **American Robin** (*Turdus migratorius*): Common to abundant year-round at lower elevations and from April through October at higher elevations. Because it is a short-distance migrant, there may be brief gaps between wintering and summering populations in some areas.
- [ ] **Gray Catbird** (*Dumetella carolinensis*): Occasional in dense riparian habitat between May and mid-September. Locations include Fred Baca Park and Monastery Lake.
- [ ] **Curve-billed Thrasher** (*Toxostoma curvirostre*): Uncommon year-round in upland desert scrub habitat, including Cochiti Lake area, La Cieneguilla Marsh, and RDAC.
- [ ] **Sage Thrasher** (*Oreoscoptes montanus*): Uncommon in summer in Great Basin scrub habitat on the Taos Plateau, including near the Rio Grande Gorge Bridge and Rest Stop. Migrates south in September and often seen along FR 151 and Tetilla Peak Road.
- [ ] **Northern Mockingbird** (*Mimus polyglottos*): Common between mid-April and mid-September in upland desert scrub, piñon-juniper, and grassland habitats.
- [ ] **European Starling** (*Sturnus vulgaris*): Common to abundant in urban and agricultural areas. Tends to be more prevalent during winter, when large flocks feed and roost together.
- [ ] **American Pipit** (*Anthus rubescens*): Uncommon breeder at alpine tundra locations, including Williams Lake above Taos Ski Valley and Santa

Sage Thrasher

Barbara Ridge. Generally not found in winter north of Cochiti Lake, where it is occasional.

- [ ] **Cedar Waxwing** (*Bombycilla cedrorum*): Uncommon migrant along the Rio Grande, casual winter visitor at Villanueva SP, and summer resident and breeder in Chama.
- [ ] **Lapland Longspur** (*Calcarius lapponicus*): Casual late fall and winter migrant and visitor along the eastern plains near Springer Lake and Maxwell NWR.
- [ ] **Chestnut-collared Longspur** (*Calcarius ornatus*): Occasional migrant and winter visitor between late September and March at sites along I-25.
- [ ] **McCown's Longspur*** (*Rhynchophanes mccownii*): Casual migrant at Maxwell NWR and Springer Lake.
- [ ] **Northern Waterthrush** (*Parkesia noveboracensis*): Casual spring and fall migrant in riparian habitats.
- [ ] **Orange-crowned Warbler** (*Oreothlypis celata*): Uncommon to com-

mon migrant and uncommon summer resident at high-altitude mixed conifer locations, including Taos Ski Valley and Santa Fe Ski Area.

- [ ] **Nashville Warbler** (*Oreothlypis ruficapilla*): Casual spring and fall migrant, particularly in Santa Fe/Los Alamos area.
- [ ] **Virginia's Warbler*** (*Oreothlypis virginiae*): Uncommon in riparian-related piñon-juniper and ponderosa pine habitats, including upper Pecos Canyon and Rio Chama Wild River Area between late April and mid-September.
- [ ] **MacGillivray's Warbler** (*Geothlypis tolmiei*): Uncommon migrant at a variety of riparian locations and occasional summer resident in the Jemez Mountains, May through mid-September.
- [ ] **Common Yellowthroat** (*Geothlypis trichas*): Locally common in cattail marshes, including La Cieneguilla, OOFL, and Fred Baca Park, from mid-April through September.
- [ ] **Yellow Warbler** (*Setophaga petechia*): Common migrant and summer resident in riparian habitat between April and September.
- [ ] **Yellow-rumped Warbler** (*Setophaga coronata*): The "Audubon's" race (*S. c. auduboni*) is a common summer resident and breeder in ponderosa pine and mixed conifer forests. Both races, "Myrtle" (*S. c. coronata*) and "Audubon's," are common winter residents and migrants at lowland sites.
- [ ] **Grace's Warbler*** (*Setophaga graciae*): Common summer resident and breeder in ponderosa pine habitat in the Jemez and Sangre de Cristo Mountains, arriving mid-April and leaving by mid-September.
- [ ] **Black-throated Gray Warbler** (*Setophaga nigrescens*): Uncommon summer resident and breeder in piñon-juniper habitat, including RDAC and Bandelier NM, between mid-April and August.
- [ ] **Townsend's Warbler** (*Setophaga townsendi*): Uncommon fall migrant throughout the area from mid-August through the end of September.
- [ ] **Wilson's Warbler** (*Cardellina pusilla*): Common spring and fall migrant throughout north-central New Mexico and occasional high-altitude breeder in the Sangre de Cristo Mountains.
- [ ] **Yellow-breasted Chat** (*Icteria virens*): Locally common in cattail marshes, including La Cieneguilla, OOFL, and Fred Baca Park, from May until September.
- [ ] **Green-tailed Towhee** (*Pipilo chlorurus*): Uncommon summer

resident and breeder in a variety of shrub habitats from mid- to high elevations, between mid-April and late September.

- ☐ **Spotted Towhee** (*Pipilo maculatus*): Abundant year-round resident and breeder in habitats with dense shrubs and abundant leaf litter. Migrates to lower elevations for the winter.
- ☐ **Rufous-crowned Sparrow** (*Aimophila ruficeps*): Casual in rocky hillsides with upland desert scrub habitat.
- ☐ **Canyon Towhee** (*Melozone fusca*): Common year-round in upland desert scrub and piñon-juniper habitats, including Bandelier NM, Cochiti Lake, and RDAC.
- ☐ **Cassin's Sparrow** (*Peucaea cassinii*): Occasional migrant and breeder in prairie grasslands. Sings and skylarks much later in the season than other passerines. A few nest each year at Maxwell NWR.
- ☐ **American Tree Sparrow** (*Spizella arborea*): Uncommon between November and mid-March at Maxwell NWR and LVNWR.
- ☐ **Chipping Sparrow** (*Spizella passerina*): Common summer resident and migrant at a variety of habitats at sites throughout this guide from early April through early October.
- ☐ **Clay-colored Sparrow** (*Spizella pallida*): Occasional migrant seen during September in open brushy areas or grasslands.
- ☐ **Brewer's Sparrow** (*Spizella breweri*): Uncommon local resident on the Taos Plateau, including near the Rio Grande Gorge Bridge Rest Area, and an uncommon spring and fall migrant.
- ☐ **Black-chinned Sparrow*** (*Spizella atrogularis*): Casual in desert scrub and piñon-juniper habitats near canyons, including Cochiti Lake and Orilla Verde areas, between May and early September.
- ☐ **Vesper Sparrow** (*Pooecetes gramineus*): Uncommon summer resident and migrant through grassland and montane shrub habitats, including Maxwell NWR, Eagle Nest SP, and VCNP.
- ☐ **Lark Sparrow** (*Chondestes grammacus*): Uncommon to common summer resident and migrant in grassland habitats between May and September.
- ☐ **Black-throated Sparrow** (*Amphispiza bilineata*): Common breeder in desert scrub habitat. Stands of cholla cactus along Tetilla Peak Road are a good location. Arrives in early March and remains until late fall.
- ☐ **Sagebrush Sparrow** (*Artemisiospiza nevadensis*): Uncommon summer resident in Great Basin scrub habitat near and west of the Rio

Grande Gorge Bridge Rest Area and Rio Grande del Norte NM—Wild Rivers Area.

- ☐ **Lark Bunting** (*Calamospiza melanocorys*): Occasional breeder at Maxwell NWR and migrant in grasslands along I-25.
- ☐ **Savannah Sparrow** (*Passerculus sandwichensis*): Uncommon in summer in wet montane meadows such as VCNP and Eagle Nest SP. Uncommon migrant.
- ☐ **Grasshopper Sparrow** (*Ammodramus savannarum*): Occasional migrant and summer resident at Maxwell NWR.
- ☐ **Song Sparrow** (*Melospiza melodia*): Uncommon migrant through north-central New Mexico and a common winter resident at lower elevations between mid-September and April or May, with the greatest abundance occurring in December. It is also an uncommon breeder at a few locations in the Jemez Mountains and Monastery Lake.
- ☐ **Lincoln's Sparrow** (*Melospiza lincolnii*): Uncommon in subalpine meadows and montane shrub habitats such as the Santa Fe Ski Area during the summer, migrating to lower elevations during the winter.
- ☐ **White-throated Sparrow** (*Zonotrichia albicollis*): Occasional visitor during the winter in flocks of White-crowned Sparrows, but not predictably at any location. Scan each flock of White-crowned Sparrows carefully.

Sagebrush Sparrow (photo by Joe Schelling)

- ☐ **White-crowned Sparrow** (*Zonotrichia leucophrys*): Uncommon breeder above the tree line, including Santa Barbara Ridge. Common winter visitor at lower elevations.
- ☐ **Dark-eyed Junco** (*Junco hyemalis*): The "Gray-headed" race (*J. h. caniceps*) is a common summer resident and breeder at high elevations in the Jemez and Sangre de Cristo Mountains. In mid-September, migrants arrive in the lowlands, primarily "Oregon" (*J. h. oreganus*), along with "Gray-headed," "Pink-sided" (*J. h. mearnsi*), and "Slate-colored" (*J. h. hyemalis*).
- ☐ **Hepatic Tanager** (*Piranga flava*): Occasional summer resident in ponderosa pine and pine-oak woodlands, arriving in early May and leaving by the end of August. The best locations are at Bandelier NM.
- ☐ **Summer Tanager** (*Piranga rubra*): Occasional migrant and summer resident along the Rio Grande.
- ☐ **Western Tanager** (*Piranga ludoviciana*): Uncommon summer resident and breeder at moderate and high elevations in Sangre de Cristo and Jemez Mountains and uncommon migrant along the Rio Grande.
- ☐ **Rose-breasted Grosbeak** (*Pheucticus ludovicianus*): Casual spring migrant in same habitats as Black-headed Grosbeak.
- ☐ **Black-headed Grosbeak** (*Pheucticus melanocephalus*): Common summer resident and breeder in riparian up through mixed conifer habitats between early May and mid-September.
- ☐ **Blue Grosbeak** (*Passerina caerulea*): Common summer resident and breeder in riparian areas along the Rio Grande and along I-25 between early May and early October.
- ☐ **Lazuli Bunting** (*Passerina amoena*): Uncommon and variable migrant and summer resident in riparian areas with brushy understory, primarily along the Rio Grande between late April and September.
- ☐ **Indigo Bunting** (*Passerina cyanea*): Occasional spring migrant along the Rio Grande.
- ☐ **Dickcissel** (*Spiza americana*): Casual migrant and summer resident in grassland habitat at Maxwell NWR and LVNWR.
- ☐ **Red-winged Blackbird** (*Agelaius phoeniceus*): Abundant year-round resident and breeder in marshes and wet fields.
- ☐ **Eastern Meadowlark** (*Sturnella magna*): Locally common summer resident at VCNP.

- [ ] Western Meadowlark (*Sturnella neglecta*): Common year-round resident in grasslands and agricultural fields.
- [ ] Yellow-headed Blackbird (*Xanthocephalus xanthocephalus*): Occasional spring migrant and uncommon fall migrant in wetlands and agricultural areas.
- [ ] Brewer's Blackbird (*Euphagus cyanocephalus*): Often confused with Common Grackle, it breeds at a higher elevation in montane grassland near forest edges, including VCNP and Eagle Nest SP, where it is common. It moves to lower elevations during winter.
- [ ] Common Grackle (*Quiscalus quiscula*): Uncommon between May and September in open areas with forest edges and agricultural areas, primarily east of the Rio Grande.
- [ ] Great-tailed Grackle (*Quiscalus mexicanus*): Common year-round in wetlands, agricultural areas, and prairie habitats.
- [ ] Brown-headed Cowbird (*Molothrus ater*): Common summer resident and breeder in open-lowland and low-mountain areas from early April through September.
- [ ] Bullock's Oriole (*Icterus bullockii*): Common migrant and summer resident and breeder in riparian woodlands between late April and August.
- [ ] Scott's Oriole (*Icterus parisorum*): Casual in rocky canyons between mid-April and the end of August.
- [ ] Gray-crowned Rosy-Finch (*Leucosticte tephrocotis*): Occasional in winter at Taos Ski Valley, Red River, Angel Fire, and VCNP.
- [ ] Black Rosy-Finch (*Leucosticte atrata*): Occasional in winter at Taos Ski Valley, Red River, Angel Fire, and VCNP.
- [ ] Brown-capped Rosy-Finch (*Leucosticte australis*): Occasionally seen in winter at Taos Ski Valley, Red River, Angel Fire, VCNP, and Santa Fe Ski Area.
- [ ] Pine Grosbeak (*Pinicola enucleator*): Uncommon in spruce-fir habitat in the Sangre de Cristo Mountains and Apache Creek.
- [ ] House Finch (*Haemorhous mexicanus*): Abundant year-round resident and breeder in more populated areas. Uncommon to common in other areas.
- [ ] Cassin's Finch (*Haemorhous cassinii*): Uncommon primarily in the Sangre de Cristo and Jemez Mountains. More plentiful in some years than others.

- ☐ Red Crossbill (*Loxia curvirostra*): Uncommon year-round resident in roving flocks at high elevations in the Jemez and Sangre de Cristo Mountains.
- ☐ Pine Siskin (*Spinus pinus*): Common year-round resident found at higher elevations during the summer and lower elevations during winter.
- ☐ Lesser Goldfinch (*Spinus psaltria*): Uncommon to common in a variety of habitats throughout north-central New Mexico.
- ☐ American Goldfinch (*Spinus tristis*): Common migrant and winter visitor along the Rio Grande, Santa Fe River, and LVNWR. Locally common in summer in Taos and Chama.
- ☐ Evening Grosbeak (*Coccothraustes vespertinus*): Uncommon to common year-round, primarily in mixed conifer forests. It moves around to take advantage of the food supply.
- ☐ House Sparrow (*Passer domesticus*): Abundant introduced resident in populated areas.

Selected Resources

Allen, Craig D., A. K. Macalady, H. Chenchouni, D. Bachelet, N. McDowell, M. Venneti-er, T. Kitzberger, A. Rigling, D. D. Breshears, E. H. Hogg, P. Gonzales, R. Fensham, A. Zhang, J. Castro, N. Demidova, J-H. Lim, G. Allard, S. Running, A. Semerci, and N. Cobb. 2010. "A Global Overview of Drought and Heat-Induced Tree Mortality Reveals Emerging Climate Change Risks for Forests." *Forest Ecology and Management* 259:660–684. doi: 10.1016/jforec0.2009.09.001.

American Birding Association. 2014. *Code of Birding Ethics*. Colorado Springs, CO: American Birding Association. Available at http://www.aba.org.

The Birds of North America Online. 2014. Edited by A. Poole. Ithaca, NY: Cornell Laboratory of Ornithology. Available at http://bna.birds.cornell.edu/BNA/.

Cartron, J-L., ed. 2010. *Raptors of New Mexico*. Albuquerque: University of New Mexico Press.

Dobson, C., and W. Dunmire. 2007. *Mountain Wildflowers of the Southern Rockies: Revealing Their Natural History*. Albuquerque: University of New Mexico Press.

Larson, Peggy. 1977. *A Sierra Club Naturalist's Guide: The Deserts of the Southwest*. San Francisco: Sierra Club Books.

Martin, Craig, 2006. *Los Alamos Trail: Hiking, Mountain Biking, and Cross-Country Skiing*. Los Alamos, NM: All Seasons Publishing.

New Mexico Department of Game and Fish. 2006. *Comprehensive Wildlife Conservation Strategy for New Mexico*. Santa Fe: New Mexico Department of Game and Fish.

New Mexico Partners in Flight. 2007. *New Mexico Bird Conservation Plan*. Version 2.1. C. Compiled by Rustay and S. Norris. Albuquerque: New Mexico Partners in Flight.

Northern New Mexico Group of the Sierra Club. 2012. *Day Hikes in the Santa Fe Area*. Santa Fe: Northern New Mexico Group of the Sierra Club.

Price, G., ed. 2010. *The Geology of Northern New Mexico's Parks, Monuments, and Public Lands*. Socorro: New Mexico Bureau of Geology and Mineral Resources.

Sauer, J. R., J. E. Hines, J. E. Fallon, K. L. Pardieck, D. J. Ziolkowski Jr., and W. A. Link. 2012. *The North American Breeding Bird Survey, Results and Analysis 1966–2011*. Version 07.03.2013. Laurel, MD: USGS Patuxent Wildlife Research Center.

University of New Mexico, College of Pharmacy, Poison and Drug Information Center. 2014. *Venomous Creatures in New Mexico*. Albuquerque: University of New Mexico. Available at http://nmpoisoncenter.unm.edu/education/pub-ed/pp_tip_pages/venom_things.html.

Index

Photos and maps are indicated in *italic* type.